Field Manual
31-23 (ID)

Headquarters
Department of the Army
Washington, DC, 1998

Special Forces Mounted Operations
Tactics, Techniques, and Procedures

INITIAL DRAFT
3 March 1998

This draft is for review purposes only and does not represent approved Department of the Army doctrine.

DISTRIBUTION RESTRICTION STATEMENT:
Approved for public release; distribution is unlimited.

TABLE OF CONTENTS

PREFACE	vii
Chapter 1. Introduction	1-1
General	1-1
History	1-2
Special Forces Mounted Operations	1-3
Chapter 2. Mounted Detachment Organization	2-1
Desert Vehicle Mobility System	2-1
Vehicle Positions and Duties	2-1
Chapter 3. Planning Considerations and Preparations	3-1
General	3-1
Pre-Mission Considerations	3-2
Collective and Individual Training	3-2
Vehicle Preparation	3-3
Equipment and Personnel Preparation	3-4
Chapter 4. Operational Employment	4-1
Infiltration and Exfiltration	4-1
Movement and Formations	4-3
Methods of Travel	4-4
Movement Formations	4-4
Actions at Halts	4-5
Laager Sites	4-6
Immediate Action/Reaction Drills	4-8
Laager Site Reaction Drill	4-10
Patrol Ferry Missions	4-10
Communications	4-11
Chapter 5. Motorcycle Section Employment	5-1
General	5-1
Employment Concept	5-2
Movement	5-3
Reaction Drills	5-3
Equipment	5-4
Chapter 6. Operations in an NBC Environment	6-1
Fundamentals of NBC Defense	6-1
Immediate Decon	6-2
Operational Decon	6-3
Thorough Decon	6-4
Nonstandard Operational Decon	6-4
Decontaminant Options	6-5
Chapter 7. Mounted SFLE Operations	7-1
Organization	7-1
Equipment and Personnel Preparation	7-1

Chapter 8. Navigational Techniques	8-1
Navigator's Duties	8-1
Terrain Association	8-2
Dead Reckoning	8-4
Vehicle Orienteering	8-4
Celestial Navigation	8-7
Satellite Navigation	8-8
Chapter 9. Camouflage	9-1
Camouflage Theory	9-1
Camouflage Methods – Concealing Objects	9-2
Camouflage in the Desert	9-2
Camouflage Considerations	9-3
Chapter 10. Maintenance and Recovery	10-1
General	10-1
Preventive Maintenance Checks and Services	10-2
Desert Environmental Effects	10-2
Lessons Learned	10-3
Off-Road Driving	10-4
Recovery	10-5
Chapter 11. Logistics	11-1
General	11-1
Mission Support Site	11-1
Five R's	11-2
Multiple MSS Concept	11-3
Caching	11-4
Chapter 12. Military All-Terrain Vehicle	12-1
General	12-2
Organization	12-2
Employment Concept	12-3
Air Infiltration	12-3
Movement	12-3
Reaction Drills	12-3
Equipment	12-4
Appendix A. M1025A2 (GMV) and M1114 (Armored) HMMWV	A-1
Appendix B. Mission-Essential Task List	B-1
Appendix C. Mounted Detachment Training and Evaluation Outline	C-1
Appendix D. Mounted Detachment Training Program	D-1
Appendix E. Pre-Mission Checklist	E-1
Appendix F. Load List	F-1
Appendix G. Fuel Estimation Formula	G-1

Appendix H. Water Estimation Formula H-1

Appendix I. CH-47/MH-47 Internal Load Operations I-1

Appendix J. Sling Load Operations J-1

Appendix K. Motorcycle Training Program K-1

Appendix L. Post-Operations Maintenance Procedures L-1

Glossary Glossary-1

References References-1

FM 31-23 (ID)

PREFACE

Until the mid 1980's, the United States (U.S.) Army did not have a dedicated mounted special operations (SO) capability. Major General Guest (then Colonel), 5th Special Forces Group (Airborne) (SFG[A]), realized this shortcoming. He understood that traditionally dismounted Special Forces (SF) operations in desert environments were unrealistic. He authorized the formation of two detachment elements in 1984 to develop mounted doctrine and operational techniques. These detachments moved to Fort Bliss, Texas, and in the fall of 1986 started fulfilling this mission.

These detachments were equipped, at first, with M880 trucks and M151 jeeps. Later they appraised, evaluated, and accepted the high mobility multipurpose-wheeled vehicle (HMMWV) series vehicle as the interim desert mobility vehicle (DMV).

This manual, first printed by Company A, 1st Battalion, 5th SFG(A) in October 1987, was a compendium of lessons learned by this element from 1985 to 1987. Its initial intent was to provide a reference for training and using mounted SF detachments within the 5th SFG(A).

Since that time, the 5th SFG(A) updated and revised their Mounted Operations Manual in March 1992 and January 1993, incorporating lessons learned and new or improved equipment.

Through the years, the manual has remained essentially the same, yet revisions have been necessary to account for latest equipment updates such as the new ground mobility vehicle (GMV) and global positioning system (GPS) devices.

This field manual (FM) is a compendium of lessons learned by personnel at Fort Bliss, Fort Campbell, Fort Bragg, and overseas, to include Operations Desert Shield/Desert Storm, Restore Hope, and Provide Democracy. Its purpose is to serve as a reference for training and using mounted SF detachments in the desert on long-term, unassisted operations. Although written primarily for desert operations, the information in this manual also applies to any special operations forces (SOF) long-range vehicular operation.

The proponent of this publication is the United States Army John F. Kennedy Special Warfare Center and School (USAJFKSWCS). Reviewers and users of this manual should submit comments and recommended changes on Department of the Army (DA) Form 2028 to Commander, USAJFKSWCS, ATTN: AOJK-DT-SF, Fort Bragg, NC 28307-5000.

FM 31-23 (ID)

Chapter 1
Introduction

Special Forces Mounted Detachments are prepared to infiltrate and operate in low or medium intensity conflicts over terrain consisting of high deserts with rugged mountains to low deserts with sand dunes and salt marshes. The capability of these detachments to travel unassisted long distances in enemy rear areas gives the Joint Forces Commander/Commander Joint Special Operations Task Force an effective tool. Figure 1-1 shows a 10-man SF mounted detachment task-organized for a 500-mile (mi) 5-day direct action mission without resupply.

Figure 1-1. SF mounted detachment.

GENERAL

In preparing for conflicts in the desert environment, it is assumed that the distance from the forward operational base (FOB) to the area of operations (AO) is too great for dismounted infiltration. The desert-oriented SFG cannot rely solely on limited Air Force Special Operations Wing assets for which to infiltrate operational detachments into their AOs.

A major role for these detachments is to conduct direct action (DA), medium to long range special reconnaissance (SR), and unconventional warfare (UW) operations. They can also expect to do area reconnaissance missions and to conduct and support unconventional assisted recovery (UAR).

In addition to the previously mentioned standard missions, mounted teams can be used to transport other personnel and/or equipment in or out of their target area.

Another important role for these detachments is to conduct coalition support (Special Forces Liaison Element [SFLE]) and/or foreign internal defense (FID) missions with nations possessing extensive mounted capabilities.

To prepare for these roles, mounted detachments must develop the following capabilities:

- Operate and communicate over long distances.

- Operate effectively and continually in a hostile air environment that may severely limit or prevent air support.

- Operate in rugged terrain, both on and off road.

- Make on-site repairs on all equipment using the skills of the detachment members and on-board tools and parts.

HISTORY

The British Army first used mounted special warfare units in the desert for active patrolling and special missions during World War I. The Libyan Light Car Patrol was used very effectively to suppress the Sanusi rebellion. Colonel T. E. Lawrence, of Lawrence of Arabia fame, made vast use of camel-mounted elements to harass and destroy Ottoman Turkish rail and supply depots in, what is now, Saudi Arabia and Jordan. Later he used vehicular patrols with great success.

Between World Wars I and II, John Ball operated throughout the North African deserts with long-range vehicle excursions for mapping and exploring. His data and techniques were later incorporated into a manual for British officers who conducted vehicular patrols in the desert. Most of his navigation techniques are still valid today and in fact are the basis for traversing techniques across desert environments.

With the combined knowledge of mounted patrolling and techniques for navigation, the stage was set for the development of a specialized body of troops who could combine this knowledge into useful military operations.

The necessity for reliable information in the Western Desert Campaign during World War II brought about the organization of the Long Range Desert Group (LRDG) in 1941. This British-led, New Zealand-manned, U.S.-vehicle-equipped unit was credited as being the most reliable intelligence-gathering tool for General Headquarters Middle East. The LRDG operated out of patrol bases in the deep South Sahara Desert. The LRDG would launch "road watch" surveillance missions and other reconnaissance missions lasting for several weeks.

Another Long Range Desert unit operating in the Sahara during World War II was the Special Air Service (SAS). The SAS initially began operations by being ferried into the operational areas by the LRDG. Later, the SAS evolved into a deep penetration, vehicle-equipped unit capable of strike operations hundreds of miles behind enemy lines. The SAS was credited with destroying more aircraft on the ground than the Royal Air Force did in the air.

After World War II and up to the present, several countries have expanded the operational concept of using mounted SOF for desert environments. The British SAS and New Zealand SAS maintain mounted desert-oriented elements.

During Desert Shield and Desert Storm, U.S. Army SOF used mounted detachments with great success on SR and DA missions behind enemy lines. They also conducted FID and liaison missions (see Figure 1-2, page 1-3) with Kuwaiti and other Arab armor units. More specifically, they conducted terminal guidance for laser-guided munitions, reported on enemy dispositions, and gave battle damage assessments. They trained with the Kuwaiti armed forces as they rebuilt their army. They also reported "ground truth" (location and condition) to higher headquarters of the Arab coalition forces to which they were attached. Mounted SO teams were naturally suited to the nature of desert warfare and the swift pace of modern mechanized forces.

Figure 1-2. SF mounted detachment during Desert Storm.

In December 1992, elements of the 5th SFG(A) deployed to Somalia. SF mounted detachments conducted interdiction, convoy security (see Figure 1-3, page 1-4), security assessments of remote towns, security reconnaissance patrols, and liaison missions with United Nations forces.

SPECIAL FORCES MOUNTED OPERATIONS

Mounted operations provide relatively rapid and secure operational assets within the theater. Typical SF operations that can be expanded with the mobility of mounted detachments include:

- DA.

- SR.

- Combat support operations to include resupply and cache missions, patrol ferry and exfiltration, communications relay, site security roles, and assistance to evasion nets.

- Close air support (CAS) operations supporting United States Air Force (USAF) aircraft (a subset of DA).

- Guerrilla warfare (GW) operations.

FM 31-23 (ID)

Figure 1-3. Elements of the 5thSFG(A) in Somalia.

- UAR.

- FID. The mounted detachment has the ability to conduct expanded advisory assistance operations.

- SFLE missions with other countries involved in all phases of military operations.

There are several advantages to using mounted SF detachments for operations. They are—

- Compatibility. Mounted detachments can work with foreign and U.S. mechanized troops without additional vehicle assets.

- Mobility. Mounted detachments can cover long distances rapidly, diminishing the FOB's reliance on USAF aircraft for operational support.

- Air Movement. The mounted detachment can use a variety of aircraft for airlanded or airdropped operations.

- Endurance. Mounted detachments can remain in the field for extended periods without the need of being resupplied.

- Transportation. Mounted detachments can ferry specialized equipment and dismounted elements into AOs.

- Firepower. Mounted detachments can bring considerable firepower to bear on targets of opportunity or preplanned objectives using the weapons systems on their vehicles.

There are also some disadvantages to a mounted operational element. These disadvantages include:

- Vehicle maintenance. Detachment personnel need to be skilled in maintenance and repair, including depot-level maintenance procedures. Teams require additional tools and parts to sustain extended operations.

- Training. Detachment personnel will require additional training including mounted tactics, navigation techniques, maintenance and repair, and vehicle camouflage.

- Security. The amount of security offered in the desert declines with the size of the element. The number of vehicles involved in the mission, the tracks they leave, and noise and light discipline will increase the possibility of detection.

FM 31-23 (ID)

Chapter 2

Mounted Detachment Organization

The SF mounted detachment is organized to maximize the capabilities and flexibility of its equipment and detachment members. Although cross-trained in different vehicle duty positions, it is critical that each detachment member thoroughly understands his primary vehicular duty position so that the detachment can operate effectively and safely as a team.

DESERT VEHICLE MOBILITY SYSTEM

Mounted detachments are organized around four prime movers. The prime mover is called the GMV. The Desert Mobility Vehicle System (DMVS) consists of—

- GMV, modified M1025A2 HMMWV.
- Desert operations trailer (DOT).
- Desert operations motorcycle (DOM).
- GPS that is vehicle-mounted and -powered. It can also be used in a dismounted mode.
- Crew-served weapon system, caliber .50 M2 heavy barrel (HB) machine gun (MG) or Mark 19 40-mm grenade launcher (GL) MG.
- Vehicular radio communications system with frequency hopping and secure capabilities. The system also has a battery box to use the receiver-transmitter (RT) in a dismounted mode.

Optimally, a mounted detachment will have 4 GMVs, 2 DOTs, 2 DOMs, 2 M2 HBs, 2 Mark 19s, 4 GPSs, 12 night vision goggles (NVGs) (AN/PVS-7), and 4 vehicle-mounted radios.

Mounted detachments will modify their vehicles to best suit their missions and standing operating procedures (SOPs) (see Figure 2-1, page 2-2). During military operations other than war (MOOTW) in an urban environment, the mounted detachment may use the M1114 (armored) HMMWV. Appendix A provides the capabilities and comparison of these two vehicles.

VEHICLE POSITIONS AND DUTIES

The mounted detachment is divided into two sections, each consisting of 2 GMVs, 1 DOT, and 1 DOM. The two sections normally operate together as a full team, but they also move and operate as split teams. Each section has six personnel. The detachment can also operate as four separate elements for a very short time with each vehicle operating independently. Regardless of how many elements the detachment is broken into, three men man each vehicle. The duty positions are driver, navigator, and weapons system operator. Each man has to be well versed in the duties and responsibilities of all duty positions. Personnel configuration of the section will depend upon the individual skills of the detachment members. Below is an example of vehicle positions and primary duties.

Figure 2-1. Mounted Detachment with modified GMV at National Training Center (NTC).

Vehicle Positions

Vehicle #1:

- Primary driver: Engineer Supervisor.
- Weapons system operator: Weapons Sergeant (SGT).
- Navigator: Assistant Operations SGT.

Vehicle #2:

- Primary driver: Medical Supervisor.*
- Weapons system operator: Communications SGT.
- Navigator: Detachment Commander.

Vehicle #3:

- Primary driver: Medical SGT.
- Weapons system operator: Weapons Supervisor.*
- Navigator: Assistant Detachment Commander.

Vehicle #4:

- Primary driver: Communications SGT.
- Weapons system operator: Engineer SGT.
- Navigator: Operations SGT.

*DOM-1—Medical Supervisor.

*DOM-2—Weapons Supervisor.

FM 31-23 (ID)

Duty Descriptions

Primary Driver. He does the preventive maintenance checks and services (PMCS) with assistance from the vehicle crew. He assumes most of the vehicle operating duties. He ensures that the vehicle is topped off with fuel at the end of each night's movement, and that the vehicle is prepared for the next night's movement. He also monitors the fuel, water, and rations level for the vehicle. He advises the vehicle command (the navigator) of the situation before the next night's movement.

Weapons System Operator. He is responsible for that vehicle's onboard weapons system. Standard armament for a mounted detachment is one caliber .50 M2 HB MG and one Mark 19 40-mm GL MG per section (total of 2 M2 HBs and 2 Mark 19s). Usually the Mark 19 GL MGs are positioned on vehicles #2 and #3 and the M2 HBs are on vehicles #1 and #4. The weapons system operator observes for enemy activity in his vehicle's assigned sector during movement. From his position outside of and on the top of the vehicle, he has the greatest field of view and his vision is unrestricted by windows and doors. He communicates with the navigator and the driver to alert them to any hazards or obstacles in the path of the vehicle or enemy activity. The weapons system operator is accountable for the internal load of the vehicle. He ensures every day at the end of the night's movement that the internal configuration of the vehicle is squared away, that everything is secured to the vehicle, and that essential equipment is accessible. He advises the vehicle commander daily on the vehicle's weapons and ammunition status.

Navigator (Vehicle Commander). The navigator in vehicle #1 is the primary navigator for the detachment. He should be able to determine position at any time within one hundred meters with a GPS or within one-quarter mile without. The other three vehicle navigators check the primary navigator and help him negotiate obstacles. He does the route planning, to include preparing the route-planning log. He does the PMCS of the vehicle's communications system. He always makes sure that the correct frequencies and crypto keys are loaded. He ensures spare batteries are accessible in case of battery failure during movement. He maintains the GPS and the vehicle's compass. The navigator also accounts for all additional equipment that is stored in the vehicle storage bins behind the driver's seat.

Motorcycle Rider. When deployed, the motorcycle riders come from vehicles #2 and #3. Vehicles #2 and #3 are the prime movers for the DOMs and act as a "mother ship" for the motorcycles. The motorcycle riders maintain the DOMs with assistance from their vehicles' crew. When deployed, the motorcycle section, never operating as single DOMs, can scout the tentative route, reconnoiter point or area targets, and act as a forward warning element for the detachment.

Chapter 3
Planning Considerations and Preparations

Planning and preparation for a mounted mission starts long before the detachment is alerted. Preparations include training and rehearsals needed to prepare the team to move 1,000 mi or 10 days in the desert unassisted.

GENERAL

The distance from the FOB to the unconventional warfare operational area (UWOA) or operational area or even the staging (launch) site may require other transportation means than the GMV. Various combinations of aircraft, rail line, and/or surface ships may be required to get the mounted detachment positioned to infiltrate an operational area. These combinations may also be used to increase the operational range of the mounted detachment by decreasing the required distance for overland infiltration.

When an operation requires both aircraft and surface ships, or other combinations, a rendezvous must take place to transfer the operational element. The method selected should be one that will land or position the element with the least chance of detection as close as possible to its AO and as simply and rapidly as possible. Factors for consideration are—

- Security.
- Size of the element.
- Operational requirements relating to the overt or covert nature of the mission.
- Capabilities of personnel and equipment loads.
- Availability of transport and delivery capabilities.
- Weather, terrain, hydrographic, and astronomical data and conditions in the delivery area.
- Enemy and friendly situation in the delivery and operational areas.

Operational elements may be delivered into the staging area or the AO using the following modes of transportation:

- Surface ships.
- Amphibious landing craft.
- Fixed- or rotary-wing aircraft.
- Rail lines.
- Line haul transport.
- Any combination of the above.

PRE-MISSION CONSIDERATIONS

In planning for a successful infiltration, consider the following factors:

Mission. The mission determines what and how much ammunition and demolitions are necessary, including special equipment.

Enemy and Friendly Situation. Order of battle affects the routes, communications procedures and capabilities, external exfiltration capabilities, and sources of resupply.

Troops Available and Training Level of Detachment Personnel. SF operational detachments are proficient in air infiltration and dismounted operations; however, long-range mounted operations require special training. This special training includes cross-country and night driving with and without night vision aids, vehicle navigation, vehicle maintenance, recovery operations, and mounted weapons system use.

Terrain and Weather. Terrain and weather also affect the route planning, personal equipment, and special equipment needs. Light conditions determine the time available for movement with regard to the enemy situation.

Time and Distance. These factors primarily affect the amount of required fuel and subsistence for detachment members, since distance and duration are similar.

Civilian Populace. Mission planning must consider the local civilians in the AO and what to do in case of mission compromise.

Equipment and Supplies. The previous six considerations determine the detachment's logistical needs. The detachment must plan for the minimum levels of all needed supplies. Mission-essential equipment and supplies will prioritize available space. During planning, the detachment may find that pre-positioned equipment is available in the AO. This equipment can range from fuel and water to a complete GMV with weapons, communications equipment, and prescribed load list (PLL). Pre-positioned supplies greatly reduce the amount of vehicles and equipment the detachment must deploy with overseas and generally speeds up their overall deployment. However, when planning for such equipment, the detachment must allocate time to inspect and prepare the equipment when it arrives in country.

COLLECTIVE AND INDIVIDUAL TRAINING

Mounted detachments are as specialized as a SCUBA, mountain, special operations techniques (SOT), or high altitude low opening (HALO) detachments. All require specialized training to become proficient.

Collective Training. Training required for the mounted detachment includes cross-country and night driving with and without night vision aids, vehicle navigation, vehicle infiltration, mission support site (MSS) and hide site establishment, vehicle maintenance, recovery operations, mounted battle drills, and dismounted crew battle drills. Priority on detachment collective training for the vehicles must always be in maintenance. The team members have only each other to depend on when deep inside enemy territory and they can never know enough about working on their vehicles. See Appendix B, example of a mission-essential task list and Appendix C, example task summaries for a mounted detachment. These two examples provide the commander a tool to evaluate his detachment.

Appendix D provides an example of a mounted detachment's training program that the 5th SFG(A) used to train elements of the 20th SFG(A) at the NTC.

Individual Training. All military occupational specialties (MOSs) require special skills or knowledge to effectively augment the mounted detachment. Detachment members that have a mechanized background are always an asset. The following paragraphs address individual training for the—

- Detachment commander—mounted mission planning, detachment mounted training concepts, mounted employment, and battle drills.

- Assistant detachment commander—mounted mission planning, long-range training planning, mounted employment, and battle drills.

- Operations sergeant—mounted mission planning, short-range training planning and implementation, mounted employment, battle drills, vehicle maintenance, and load planning.

- Assistant operations sergeant—mounted mission planning, mounted employment, battle drills, and vehicle maintenance.

- Weapon sergeants—mounted employment, battle drills, and specialized weapons.

- Engineer sergeants—mounted employment, battle drills, specialized weapons, and load planning.

- Communication sergeants—mounted employment, battle drills, specialized electrical wiring techniques, and vehicle maintenance.

- Medical sergeants—mounted employment, battle drills, specialized medical techniques, and vehicle maintenance.

Cross-Training. Mounted detachment personnel require thorough cross-training. Each vehicle must be able to operate independently for extended periods. Place priority on communications, medical training, basic employment, and maintenance. Skills not practiced are skills lost. Do not wait until isolation to cross-train.

VEHICLE PREPARATION

Detachment personnel make all preparations necessary for airlanded, paradrop, seaborne, and overland insertions. They must plan for and spend sufficient time to prepare their vehicles for the assigned mission, from infiltration to exfiltration. There are no motor pools in the AO where the detachment can effect repairs; all maintenance and repair operations take place in the field. The GMV is not only a mode of infiltration and exfiltration; it also is a duration and distance enhancement and survival platform for the mounted detachment.

Detachment members load each vehicle so that it can act independently during the mission. They must carefully consider weight. Too much equipment is just as bad as not enough. An overloaded vehicle handles poorly, consumes fuel at a higher rate, lacks power, and will experience more maintenance problems. Items having the greatest effect on weight are fuel, water (50 pounds [lb] per 5-gallon [gal] container), ammunition by type (including shipping containers), and personal equipment. Try to limit carrying unnecessary equipment. There is a tendency to carry more equipment because there is room. Knowing one's vehicle greatly enhances mission success and the crew's trust in their vehicle. Avoid borrowing or loaning vehicles at the last moment. If you must borrow or loan, allow enough time for detachment members and motor pool personnel to perform pre-mission maintenance. Think of these vehicles as used cars. After buying a used car, you would not immediately go on a 1,000-mi trip without proper maintenance. A common mistake is to assume that all GMVs are the same. Although they may look the same on the outside, each one performs differently based on age, mileage, past maintenance, and hours on the engine.

EQUIPMENT AND PERSONNEL PREPARATION

An important aspect to pre-mission preparation is vehicle maintenance and keeping all equipment in a go-to-war status. Members must inspect and exercise their vehicles even while in garrison (see Figure 3-1). The detachment operations sergeant is responsible for status of the detachment's vehicles. The vehicle navigator is responsible for the status of his vehicle.

Ensure proper PMCS in garrison by moving all vehicles out of the motor pool monthly to exercise and test the equipment. This test should include on- and off-road operation in all gears. Check for wheel alignment and listen for any unusual noises. A vehicle left alone in the motor pool will break down. The more these vehicles are exercised, the better they will work.

Figure 3-1. PMCS at motor pool.

Keep the basic equipment common to each mission on the vehicle at all times (see Figure 3-2, page 3-6). This equipment includes tools, petroleum, oils, and lubricants (POL), spare parts, recovery items, tire repair kits, and other miscellaneous items. Such actions will not only save loading time and storage space needed to store these items between missions, but they reduce the chance that these items will be forgotten.

Prepare each vehicle using a vehicle loading list. This list is compiled from team SOPs, experience, and mission requirements. Simplicity is the key to success. A good tool is a vehicle loading plan that standardizes the location of equipment common to all in each vehicle. This plan ensures that anyone assigned to the detachment can go to any vehicle and locate or pack team equipment.

Control and assist the preparations after alert using pre-mission checklists (see Appendix E). The detachment operations sergeant ensures the completion of the pre-mission requirements.

Conduct inspections to ensure the vehicles are loaded properly. Appendix F contains an example load list without regard to specific mission requirements, other than a planning figure of 10 days or 1,000 mi.

Upon receipt of a notice to deploy, inspect the detachment vehicles as soon as possible to ensure mechanical reliability. Conduct this inspection at least 30 days before vehicle shipment (or as early as possible) to allow motor pool personnel time to correct deficiencies. Do not inspect the vehicles only per the operator's manual—conduct a very thorough going-over from top to bottom. A good reference to follow for this inspection is the annual inspection required for the HMMWV. Motor pool personnel will help inexperienced detachment personnel perform this inspection. It is key that the detachment personnel be present at this inspection. Your life may depend on your vehicle. Test-drive each vehicle to ensure mechanical reliability. Make sure this inspection takes the vehicle up to operating temperatures. Also, check climbing ability, winch operation with load, transmission and transfer case performance through all gears on challenging terrain, engine performance, front and rear wheel alignment, and listen for any unusual noises or rattles. After this inspection and test, rate each vehicle by performance. The stronger vehicles should perform the more challenging aspects of the mission. Avoid overloading or hauling trailers with the weaker vehicles. The next inspection should take place 3 to 5 days before load out or during isolation. Inspect the items normally kept on the vehicle and all mission-related equipment. A good way to inspect this equipment is to separate the mission-essential equipment by vehicle. Each vehicle team inspects its own equipment to ensure reliability of one's own equipment and ability to operate the equipment. The last inspection should be the normal final inspection or spot check done during the last few hours before the infiltration or shipment of the equipment.

Plan for sufficient fuel supplies. Fuel trucks or fuel points are not available in the mission area. Frequently, it is difficult or impossible to get any kind of resupply. A general planning figure is nine miles per gallon (mpg) for initial estimation of fuel requirements. Use the formula in Appendix G to plan for fuel usage.

Plan for and take adequate water. Minimum water planning figures are four to six quarts per man per day for mounted operations in the desert. Take additional water for dismounted missions within the mounted role. Do not count the water carried on individual load-bearing equipment (LBE) for this requirement. Detachment members use a vehicle water bottle for the crew. They never use the water supplies on their LBE unless separated from the vehicles during dismounted operations or when placed in a survival or evasion situation. As a rule, consume water from the vehicle's stores first before using personal stores. Use the formula in Appendix H to plan for water usage.

Plan for and take adequate food supplies. Remember that food consumption in hot, dry climates is generally less than in other climates. Individuals should pack the majority of their food items in a food bag (ditty bag) instead of their rucksack to limit the extent of unpacking their rucksack when getting meals. A ditty bag ensures they will have a minimal kit of food and survival and evasion items on hand. Construct the ditty bag from a durable bag large enough to hold 3 days of food, minimum sleeping gear, personal escape and resistance gear, first-aid kit, and personal toilet articles. Pack a minimum of three meals in the rucksack, so that the detachment member will have a food supply if required to abandon the vehicle rapidly. If several cases of food are packed on the vehicle, the crew avoids opening more than one case at a time. This action helps when estimating the duration of remaining food and cuts down the constant shuffling of equipment.

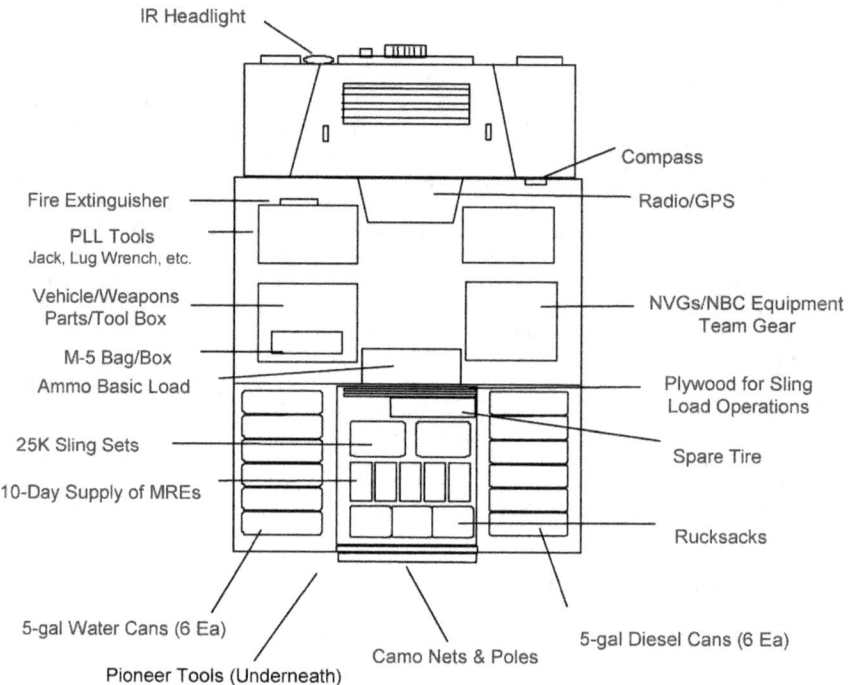

Figure 3-2. Example vehicle load configuration.

Place ammunition where it can be accessed quickly. Secure large ammunition cans or containers to prevent injury in accidents due to shifting loads. Carry a small basic load of demolitions separately to deal with contingencies (duds or mines). Construct and position a vehicle destruction kit for quick accessibility. Each member should have three basic loads of small-arms ammunition: one on the LBE (primary), one in rucksack (alternate), and one in an ammo can positioned in the vehicle (contingency). The ammo can in the vehicle should contain all contingency ammunition for the crew. Position basic signaling ammunitions near the navigator's position. These would include colored smokes and colored starclusters to aid in identification.

Plan for maintenance and repair contingencies based on the mission, the terrain and weather in the operational area, mission duration, and maintenance experience. The mounted detachment normally carries one general mechanic's toolbox with metric supplement per section. Additionally, each vehicle carries its own operator vehicle maintenance (OVM) set. Each vehicle also carries a small supply of motor oil (15w-50), Dextron II transmission fluid, and brake fluid for basic maintenance needs. For long-duration missions, the trailer towed by the second and third vehicles carries the majority of the maintenance supplies. Each vehicle should also carry one complete replacement set of fluids, including motor oil, transmission fluid, brake fluid, and antifreeze. Carry basic spare parts such as fan belt, upper and lower radiator hoses, and main fuel tank drain plug. Construct a general repair can to carry such items as tire plug kit, automotive liquid metal, assorted hose clamps, and radiator repair kit. On long-duration missions requiring trailer usage, construct an additional spare parts box to carry such items as starter, alternator, half shafts, glow plugs, and battery. The detachment will normally carry enough POL and PLL to repair or replace any maintenance problem in the field if it is at all possible to repair or replace. See Appendix F for a recommended list of spare parts for a generic mission of 1,000 mi or 10 days.

Once everything is packed and ready for deployment, strap down and secure all equipment and supplies against movement inside the vehicle. Cross-country driving makes it essential that all equipment be tied down securely.

Chapter 4
Operational Employment

The success of the mission and survival of the operational detachment lies in its ability to infiltrate, move, conduct operations, and exfiltrate—all without being detected. In mounted operations, survival depends upon moving solely at night and using proper camouflage measures during the day.

INFILTRATION AND EXFILTRATION

The threat to each method of infiltration and exfiltration is different. The following paragraphs illustrate typical threats to a mounted detachment when infiltrating by air or by ground.

Air Infiltration and Exfiltration

Mounted detachments infiltrating and exfiltrating by air must avoid an extensive and integrated enemy air defense system. Such a system provides complete coverage at medium to high altitudes with a high redundancy of coverage in heavily defended areas. Soviet doctrine, currently used by many nations in the Middle East, has made concerted efforts to improve low-altitude detection.

Ground Infiltration and Exfiltration

Mounted detachments infiltrating and exfiltrating by land must avoid hostile border security forces. These forces employ sensors, minefields, other barriers, patrols, checkpoints, and other populace control measures to detect clandestine movement across closed borders. Once the mounted detachment crosses the border, it still faces rear area security threats.

Planning Considerations

The following paragraphs address the planning considerations for airborne and ground infiltration.

Airborne Infiltration. The mounted detachment can use several platforms to infiltrate its mission area.

C-130/MC-130. The C-130 Hercules aircraft has a great deal of advantages as an infiltration platform. Some planning considerations are—

- The team can fit two vehicles per aircraft.

- Weapons system will be mounted and cleared.

- Vehicle will be mission ready with the exception of ammunition in the weapons system.

- Everyone will ride on the aircraft.

- Fuel tanks have to be half empty on C-130 aircraft, without waiver. MC-130s will normally allow the vehicles on with a full tank but full tanks must be coordinated beforehand.

FM 31-23 (ID)

- Need a C-130 capable dirt strip (916 meters)

MH-47 Helicopter, Internal Load (see Appendix I). The GMV will fit inside a CH-47 or MH-47 helicopter with two inches of clearance around the vehicle (see Figure 4-1). This clearance makes for a very tight fit and must be carefully rehearsed with the aircrew. Planning considerations for this aircraft are—

- Rigging the vehicle.

- No objects extending from the top or sides of the vehicle.

- The weapon system will be stored as one unit.

- Cannot load with trailers.

- Rehearsal time with driver and aircrew.

- Landing zone (LZ) or pickup zone (PZ) must be flat. Any surface undulation will cause the internal frame of the Chinook to bend. This bend will lock the GMV in the helicopter or prevent it from being loaded.

Figure 4-1. Loading GMV in an MH-47 for infiltration.

MH-47 Helicopter, Sling Load (see Appendix J). Using procedures developed with 5th SFG(A) and Task Force 160, the MH-47 can land, hook up the vehicle, and load the vehicle crew on the same aircraft. The procedures for working with an MH-47 are different from conventional sling load operations and require coordination and rehearsals. Planning considerations include—

- Additional sling sets needed.
- Rigging the vehicle.
- Trailer(s) cannot be sling loaded.
- Rehearsal with aircrew required.

Ground Infiltration and Exfiltration. The HMMWV leaves a unique vehicle signature that makes it difficult to conceal its tracks. Take extreme care during route selection. Other planning considerations are—

- *Rigging Vehicle.* A common mistake is to take everything except the kitchen sink when using the GMV. Take care to properly load and configure the vehicles for a long distance movement.
- *Trailer(s).* These can be taken for use en route or cached.

MOVEMENT AND FORMATIONS

When planning and conducting movement, consider the below listed fundamentals of movement to reduce chance of enemy observation and contact.

Cover and Concealment

Use terrain features and vegetation that offer protection from enemy observation. When using cover and concealment to its full advantage, a trade-off usually exists between security and speed of movement.

Skylining

Avoid skylining. Select routes that avoid high ground that may silhouette the vehicles.

Chokepoints

Avoid chokepoints. Chokepoints or areas where the terrain naturally channels routes are often sites for ambushes or areas that the enemy may have under observation. If a chokepoint proves impossible to avoid, then reconnoiter it thoroughly before moving through it.

Populated Areas

Avoid known or suspected populated areas. In the Middle East, this means all water holes because the populace and therefore the enemy know all water holes. A mounted detachment cannot move covertly if people know they are in the area.

Movement Discipline

Practice movement discipline. Movement discipline means adhering to your light, noise, litter, and interval rules. It also means keeping your speed slow enough so that you do not leave a large dust signature (usually 10 to 12 miles per hour (mph) on most surfaces at night, slower during the day).

Security

Maintain 360-degree security at all times to avoid being taken by surprise. The detachment operations sergeant and/or the unit SOP assigns a sector of fire and observation to each vehicle during movement and at halts.

Routes and Contingencies

Make sure all detachment members know the route and contingency plans.

METHODS OF TRAVEL

There are two methods of travel in the operational area. They are either on existing tracks, trails, or roads, or traveling off-road or cross-country. There are advantages and disadvantages to both.

Trails/Tracks

Advantages are speed of movement, hard packed trails do not easily yield readable prints and signs of passage, quietness of movement, less stress on vehicles and tires, and navigation is sometimes easier.

Disadvantages are usually a greater chance of being seen or compromised, natural lanes of observation and fire exist for the enemy, and mechanical and/or manual ambushes are more probable. The U.S. HMMWV (the platform used by the GMV) leaves a distinctive tire trail unlike any other truck. Consider this fact during planning.

Cross-Country

Advantages in traveling off-road are there is less chance of enemy observation or contact, usually afford more cover and concealment, and there is less chance of an ambush.

Disadvantages are slower rates of movement, more noticeable vehicle tracks and signs of passage, tire failure and vehicle stress is greater, and navigation is usually more difficult. Some desert terrain is so rough that even the GMV has trouble traversing it faster than a man can walk. It is vital that the detachment rehearses cross-country movement in terrain as close as possible to that of the target area before deployment.

MOVEMENT FORMATIONS

The mounted detachment can employ five movement formations to suit the situation. These are—

Traveling Column

Use this formation when contact is not likely. Use the visibility rule for interval. Illumination conditions, terrain and vegetation, and night vision equipment affect this rule. The driver keeps the vehicle to his front in sight.

Traveling Overwatch Column

Use this formation when enemy contact is possible but not probable. The driver of the second vehicle increases his interval from the lead vehicle. This action allows the detachment to use the rule of

FM 31-23 (ID)

making contact with the smallest element possible, allowing the remainder of the detachment to fire and move in support of the lead vehicle.

Bounding Overwatch

Use this formation when enemy contact is expected or used in retrograde when the detachment is breaking contact. Each section bounds as a team, never exceeding half of the onboard weapons system range of the section in overwatch, about 900 to 1,000 meters. The sections in overwatch provides covering fire for the bounding section. The bounding section should attempt to place itself in a position within the line of sight of the section in overwatch.

Wedge Formation

Use this formation to move through enemy positions by fighting through them when breaking contact is not feasible. This formation can also be used with extremely wide intervals, determined by visibility, to conduct search operations (see Figure 4-2).

Figure 4- 2. GMVs in wedge formation.

Diamond Formation

Use this formation when crossing extremely large open areas. Each section forms a side of the box when moving forward. Visibility determines the interval between vehicles in each section. The interval between sections should not be greater than 900 to 1,000 meters. This formation is hard to control; therefore the sections plan for and designate rally points before they separate.

ACTIONS AT HALTS

Any time the detachment conducts a planned halt (short or long), it will conduct a coordinated shutdown of all vehicles. The commander or operations sergeant initiates the shutdown using hand and arm signals. He exits his vehicle and stands where he can be seen by all the vehicles. He then waves his arm in a circle over his head and drops it toward the ground to signal all vehicles to shut down their engines at the same time. He uses the same procedure, when the halt is over, to start their engines at the same time. If it is not possible for the commander or operations sergeant to visually signal all the vehicles at the same time, he can use the radio to indicate engine shutdown or engine on.

Use of the radio should be avoided to lessen the detachment's radio signature, but it can be conducted safely if done properly.

Once the vehicles have been shut down, the detachment conducts a security listening halt before any other functions take place. The length of time for the halts will be established in planning and/or by detachment SOP.

Short-duration halts are used to communicate with higher headquarters, make necessary repairs, or establish a position fix. For halts of less than 15 minutes, the detachment does not break travel formation. Personnel man all vehicle weapons, and establish 360-degree security. For halts of longer than 15 minutes, the detachment, if possible, moves off its direction of travel and establishes one of the following positions:

- **Coil formation.** Use this formation when moving in a column formation or along a road/trail. The detachment moves into a partial perimeter along the route of march. Members of each vehicle observe their assigned section of the perimeter. The terrain determines vehicle interval, but it is not usually less than 50 meters. During the halt, necessary tasks will be performed, each man is briefed on the present location, and a contingency plan is issued if contingencies change.

- **Diamond formation.** Use this formation when moving cross-country or in a wedge/diamond formation. The detachment moves into a perimeter. Members of each vehicle observe one-quarter of the perimeter. The terrain determines vehicle interval, but it is not usually less than 50 meters. During the halt, necessary tasks will be performed, each man is briefed on the present location, and a contingency plan is issued if contingencies change.

LAAGER SITES

Laager sites or remain all day (RAD) sites are vehicular patrol bases where mounted detachments can maintain their vehicles, rest their crews, plan missions, and hide during daylight. There are two types of laager sites: short duration (occupied for only one period of daylight) or long duration (occupied for longer than one period of daylight).

During route planning, select tentative primary and alternate laager sites on the primary and alternate routes. The detachment should arrive in the general area of the laager sites about two hours before morning nautical twilight. This arrival time will allow enough time for a proper recon of the area and to emplace and camouflage the vehicles before first light.

Upon reaching a tentative laager site, or before first light, the motorcycle element or a dismounted element can reconnoiter it. Once selected, the detachment operations sergeant and primary navigator enter the site on foot and direct the incoming vehicles into position. As each vehicle is placed into position, its members are assigned their area of responsibility. After the detachment is in place, it conducts a listening period to determine if there is any activity in the area.

Tasks, in order of priority, after the listening period are—

- Ensure 100% security.

- Launch a dismounted patrol to erase vehicle signs into the laager site for a predetermined distance set by the detachment commander.

- Camouflage vehicles (one per section, the other provides security).

- Confirm sectors of fire and prepare range cards as necessary.

- Establish observation posts (OPs) or listening posts (LPs), if necessary.

- Establish field telephone communications to each vehicle.

- Reduce security, refuel, perform maintenance, and attend to personal hygiene.

The laager site does not necessarily resemble a circle. The terrain and vegetation play a role in locating each vehicle. All four vehicles may be placed in the perimeter if necessary, but normally the detachment commander's vehicle (number two) is located in the center of the laager site. This formation resembles a triangle and allows a greater arc of fire if attacked.

When selecting and preparing an SF mounted detachment's laager site, the priority is concealment, remaining undetected, and if compromised, breaking contact rapidly, not to fight the enemy and hold terrain. The detachment camouflages and positions its vehicles with this thought in mind (see Figure 4-3).

Figure 4-3. Camouflaged GMV.

The detachment may have to occupy the laager site for more than one period of daylight. Such an occupation is most common when the detachment needs to wait for more advantageous weather or light conditions before moving, has deployed a dismounted element on a mission and must remain in the area, or in a situation where extensive repairs must be made before resuming the mission. When occupied for more than one period of daylight, additional tasks include—

- Enhancing early warning measures.

- Improving continuously defensive positions (to include defensive minefields as necessary).

- Conducting recons and establishing surveillance of the area.

Upon vacating the laager site, the detachment sterilizes the site as much as possible to deny the enemy intelligence on the detachment laager site or its operations.

Terrain limitations may not allow positioning of the detachment with multiple bug-out routes and still properly conceal the vehicles. Give priority to concealing the detachment, even if it reduces its potential evacuation routes.

IMMEDIATE ACTION/REACTION DRILLS

In a worst case scenario, the detachment will find the enemy at a time and place that is most advantageous to the enemy. To counter this threat, the detachment moves at night using routes that will allow the best chance to remain undetected. Despite these precautions, the detachment must be prepared should it make contact with the enemy. It prepares itself for contact by keeping the weapons systems manned, keeping vehicle interval, and maintaining movement discipline. The detachment will rely on making contact with the smallest element (one vehicle). This action allows the rest of the detachment to fire and move in support of the lead vehicle.

The detachment can increase its ability to avoid compromise by using vehicle-mounted thermal imagers during halts and individual NVGs during movement. Without stabilizers or gyroscopes, the long-range thermal imagers are normally ineffective during movement. Use infrared (IR) lights only when necessary. More and more countries have IR capabilities and the IR headlight shows up like a spotlight under IR.

Making contact at night, even under the best of illumination, makes it difficult to determine the number of enemy involved. During unexpected enemy contact, the detachment seeks to break contact and place as much distance between itself and the enemy as the terrain and light conditions allow. Detachment SOP and experience will establish immediate action drills (IADs). Generally, the most effective way to break contact is to bound away from the enemy in pairs. Other methods include—

- **Contact from the front or rear**(Figure 4-4, page 4-9). Normally the lead or tail vehicle will make contact first. The contacting vehicle will immediately engage the enemy; the other three vehicles will move to the sides in the direction of movement and engage the enemy. The contacting vehicle will maneuver in the opposite direction passing through the detachment. As the contacting vehicle moves past, each vehicle will engage the enemy then maneuver and follow; the last vehicle will continue to engage the enemy enhancing the break of contact. The last vehicle will also deploy smoke grenades to hinder the enemy's night vision. The tail vehicle may employ pursuit deterrent devices such as M15 antitank (AT) mines and pursuit deterrent mines (PDMs).

- **Contact from the flank near and far.**The detachment is not designed to engage in decisive firefights with the enemy, so again breaking contact is desirable. The detachment must use the mobility and speed of the GMV in moving to avoid observation and therefore enemy fire.

    ◊ *Far contact.* Upon contact from the flank (Figure 4-5, page 4-9), all weapons systems will engage the enemy with as much fire as possible. The vehicle in contact will maneuver in the opposite direction passing through the detachment. As the contacting vehicle moves past the detachment, each vehicle will engage the enemy then maneuver and follow. The last vehicle will continue to engage the enemy *enhancing* the break of contact. The last vehicle will also deploy smoke grenades to hinder the enemy's night vision. The tail vehicle may use pursuit deterrent devices such as M15 AT mines and PDMs.

◊ ***Near contact.*** Upon contact from the flank when the enemy is too close to break contact, the detachment will turn into the enemy and attempt to fight their way through with all weapons available. The detachment will move by split team and link up at the last en route rally point.

- **Recovery of personnel.** During all IADs, the detachment will try to recover personnel from a down or disabled vehicle. The vehicle closest to the disabled vehicle attempts the recovery. The rest of the detachment maneuvers to provide support for the recovery vehicle.

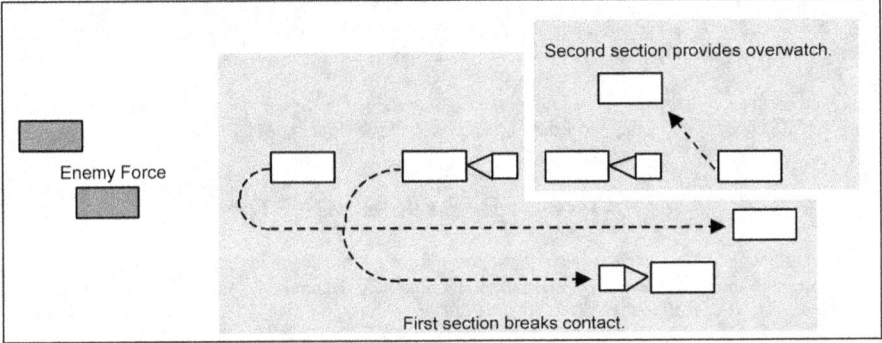

Figure 4- 4. Contact from the front.

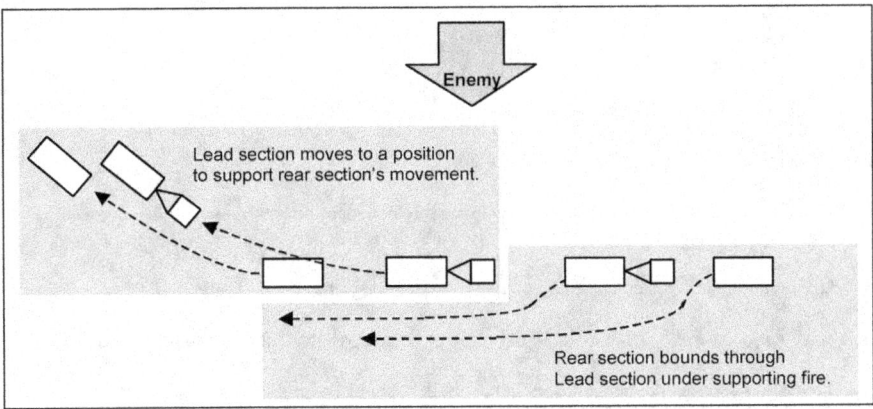

Figure 4- 5. Far contact.

On order of the detachment commander, the detachment will be prepared to split up by split team (preferred) or individual vehicle and move to any designated rally point by different routes to break contact.

LAAGER SITE REACTION DRILL

When initially setting up in a laager site, the detachment ensures that camouflage nets are set up to facilitate rapid takedown or drive away ability. The commander/operations sergeant establish two different directions for emergency evacuation (bug-out), ensure the bug-out routes are at least 120 degrees from the laager entry site, and that all personnel are aware of the routes and any bug-out emergency or contingency plans.

When the detachment is laagered, it is in its most vulnerable position. Preparing all the vehicles to fire and maneuver rapidly from their positions with little or no loss of equipment must support actions in case of compromise of the laager site.

The detachment should be prepared to vacate the laager in a violent but orderly manner at all times. Detachment members always store equipment that is not being used.

The diamond formation is the most used method to vacate the laager hastily in case of enemy attack. The lead vehicle chooses the route and leads the other vehicles, while the other three vehicles engage enemy targets.

PATROL FERRY MISSIONS

Mounted detachments can be used to move personnel and/or equipment in and out of the operational area. After moving the personnel and/or equipment, the mounted detachment can linger in the AO to support the advanced operational base (AOB) or FOB in—

- UAR.

- Exfiltration.

- Airborne operations.

- Establishing caches for future operations.

Planning considerations for ferry missions include—

- Isolating the mounted element with the dismounted element to preclude any difficulties in planning the mission to include routes, procedures, reaction drills, and contingencies.

- Giving the mounted element and its navigator deciding authority on routes.

- Assigning control over the dismounted element being ferried to the mounted commander.

- Ensuring the mounted element does not cross the forward edge of the battle area (FEBA) multiple times with dismounted elements. Such actions increase the chance of detection.

- Not tasking the mounted element with DA missions close to the dismounted operation zone when supporting dismounted infiltrations so as not to draw attention.

FM 31-23 (ID)

COMMUNICATIONS

When occupying a laager site for an extended period, set up an internal communications net using field telephones. This net reduces the signature of foot movement and radio communications.

Use secure frequency modulation (FM), with frequency hopping on low power, communications between vehicles or between mounted and dismounted elements. Such communications will decrease the range of the radio systems used, but they will hinder the enemy's ability to detect and compromise the detachment.

Mounted and dismounted detachments can use short-range, high frequency (HF) transmissions using International Morse Code or burst devices. These transmissions increase the range of communications, but are often difficult to establish or maintain. Again, use codes and maintain brevity to prevent enemy detection.

The detachment will need to make long-range communications during its mission. If it must communicate with the FOB at night during movement, it will set up and establish a perimeter as described above and communicate as rapidly as possible. The best time to communicate with the FOB is after the detachment finishes its night movement, establishes a laager site, and camouflages all vehicles.

FM 31-23 (ID)

Chapter 5
Motorcycle Section Employment

The use of motorcycles in military applications is not new. With the advent of light forces and mounted reconnaissance teams, motorcycles have proved useful as advance scout elements for mounted elements. U.S. SOF, British, and Australian SAS employ motorcycles in their mobility troops (Figure 5-1).

Figure 5-1. 5th SFG(A) Operating motorcycles in Nevada.

GENERAL

The motorcycle element provides the detachment a highly mobile and rapid capability to do—

- Route reconnaissance. It provides early warning and reconnoiters questionable sections of the intended route.

- Area reconnaissance. It reconnoiters small or large areas rapidly.

- Point reconnaissance. It locates surveillance sites, laager sites, or communications sites.

- Surveys of contaminated areas. It determines the extent of nuclear, biological, and chemical (NBC) contamination.

5-1

- Transportation tasks. It transports small amounts of equipment or supplies to distant OPs, LPs, or surveillance sites. It emplaces caches or moves personnel to communications or contact sites.

The motorcycle provides the mounted detachment the following advantages:

- Mobility. It provides excellent cross-country mobility, virtually only limited by the skill of the rider.

- Size. It is small and easy to camouflage.

- Weight. It is relatively lightweight, requiring only two people to load it onto the trailer.

- Fuel economy. It consumes minimal fuel.

- Speed. It is extremely quick and can outrun other combat vehicles if necessary.

Disadvantages to operating motorcycles in SF mounted operations include—

- Training. The off-road military motorcycle rider requires complete and detailed training in operating and maintaining the motorcycle. This training is extensive and generally much more comprehensive than what is required for a standard civilian or military motorcycle license. Appendix K contains a recommended program of instruction for SF military motorcycle riders.

- Range. The motorcycles have a limited range due to their small fuel tanks.

- Vulnerability. The rider is vulnerable to man-made and natural hazards.

- Navigation. The rider must stop to determine position. It is dangerous trying to read the map while riding.

EMPLOYMENT CONCEPT

The motorcycle section is a capability. Unless the situation dictates its use, it is not used constantly. Each section controls, transports, and provides the riders (primary and alternate) for each motorcycle. When deployed, the motorcycle section is made up of one man from the vehicle #2 and one man from the vehicle #3. This leaves three men in the lead and tail vehicle, and two men in the middle vehicles. The motorcycle section has two rules that it will never violate—

- The motorcycle section never operates as a single motorcycle.

- When the motorcycle section returns to the detachment, the first task is to refuel the motorcycles and perform PMCS. The supporting vehicles' crew refuels the motorcycle, while the riders report to the detachment commander. The riders, however, perform the PMCS.

The motorcycle section deploys ahead of the detachment at a distance determined by the terrain and the situation. The interval between the motorcycle section and the main element should be no greater than the signaling distance of the primary signaling device (usually pen flares). The detachment will establish rally points and rendezvous points with the motorcycle section before it deploys. The motorcycle section should never be farther away from the detachment than half the trip capacity of the fuel tanks. Such prevention methods ensure they can make it back to the detachment's last known site.

Riding at night, particularly with NVGs, becomes very fatiguing. Therefore, the detachment commander and primary riders must plan for driver rotation. Operational burnout occurs between 3 to 6 hours with NVG use. During daylight driving, drivers should also be replaced after no more than 6 hours of cross-country driving.

MOVEMENT

Any type of movement begins with pre-movement planning of the routes, rally points, and navigational checkpoints. The motorcycle section uses only the first two levels of navigation—vehicle orienteering and dead reckoning (DR). Plans exist, however, for equipping the motorcycle section with a GPS. Driver's logs, addressed in the next chapter, can provide a valuable navigation aid to the motorcycle section. The rider carries all maps and logs on his person should he have to separate from his motorcycle rapidly.

The motorcycle section can use any of three movement formations. Motorcycle interval during movement is based on visibility and the situation.

- Column formation. This formation is the preferred formation. Both motorcycles travel on the same path. Interval is as far as visibility permits.

- Staggered formation. The second motorcycle travels behind and to one side of the lead motorcycle. This formation allows the lead rider to see the trail rider more easily. Again, interval is based on visibility.

- Abreast formation. This formation is used when the riders need to communicate either verbally or with arm and hand signals. This formation can be used to conceal the fact that there are two motorcycles. In this formation the motorcycles will sound like one motorcycle at a short distance and will kick up what looks like a single dust trail. This formation is mostly used when chance of enemy contact is very unlikely.

During movement, the lead rider is the navigator and the trail rider is the security man who is the primary signaler if the section is compromised or enemy contact is made. The motorcycle riders should ride in as high of a gear as possible without lugging the engines to limit noise.

REACTION DRILLS

The motorcycle section is very vulnerable to small-arms fire. It must use its mobility and speed to distance itself from the enemy if contact is made.

If it makes contact with the enemy, the motorcycle section tries to break contact by placing distance and cover between themselves and the enemy. Both riders must be aware of each other. If one motorcycle goes down, the other must gain position to support the downed rider until he can make his way to either the operational motorcycle or a covered and concealed position.

At first opportunity, the trail rider must signal the main element that contact is made. He usually uses a pen flare. If the distance between the section and the main element is too great, then he uses a star cluster. The lead rider must also have signaling devices should the trail rider become a casualty.

The riders make their way, either by motorcycle or on foot, paralleling their back trail until they link up with the main element.

The motorcycle riders should be very adept at controlled ditching, so they can effectively gain the prone position if under a heavy volume of enemy fire.

EQUIPMENT

The motorcycle riders should carry mission-essential and maintenance equipment. Listed below are mandatory items of equipment for the motorcycle section, per rider:

Individual weapon and LBE, to include ammunition, compass, first aid kit, water, strobe, flashlight, maps, Department of Transportation-approved helmet (***not a Kevlar helmet***) equipped with headsets and microphones for communications, AN/PRC-126 or other small RT (for emergency contact with main element), and pen flares and/or star clusters.

Motorcycle maintenance kit that includes fix-a-flat sealant, pliers, screwdriver, tire valve core, spoke wrench, chain tightening wrench, spare spark plug, spark plug wrench, electrical tape, and master chain link set, crescent wrench, and bug-out bag (includes food, survival, and comfort items).

Chapter 6
Operations in an NBC Environment

This chapter provides mounted SF soldiers information on operating in and recovering from an NBC environment. It serves as a quick reference to basic NBC procedures and emphasizes decontamination of personnel and equipment. The threat of NBC weapons and the level of unit NBC assets available will vary with each situation. However, the mounted detachment must be trained and prepared to operate in and recover from all possible NBC scenarios.

FUNDAMENTALS OF NBC DEFENSE

Avoidance, protection, and decontamination are the fundamentals of NBC defense. All mission analysis should be viewed in conjunction with these fundamentals. (Decontamination will expand into Restoration with the revised Joint Publication 3-11.)

Avoidance

Avoidance involves assessing the threat facing the friendly force, identifying whether friendly units are targets, understanding the field behavior of chemical and biological (CB) contamination, and locating CB and toxic industrial hazards in the AO. Avoidance addresses individual and unit measures taken to avoid or minimize CB hazards. By taking measures to avoid CB hazards, units can reduce their protective postures and decrease the likelihood and extent of decontamination required (See FM 3-3).

Protection

Protection is divided into the categories of force, collective, and individual (See FM 3-4).

Force Protection involves actions taken by commanders to reduce their units' vulnerability to an NBC attack. These actions include a vulnerability assessment, a mission-oriented protective posture (MOPP) analysis, and risk reduction.

Collective Protection addresses the use of shelters that permit the reduction of individual MOPP levels.

Individual Protection involves actions taken by the soldiers to survive and continue the mission under NBC conditions, including their use of personal protective clothing.

Decontamination

Decontamination should be considered within the context of mission, enemy, terrain, troops, time available, and civilians (METT-TC) and resources available. The different origins and forms of contamination cause different hazards. Contamination can be either solid, liquid, or gas. You must be aware that NBC hazards can be transferred between surfaces, spread on the original surface, consist of a vapor, pass out in gas form from a contaminated surface in low levels (desorption/off gas), and radiation released by radioactive dust or dirt (See FM 3-5).

The following four factors must be addressed before you decide to decontaminate:

- Lethality (see FM 3-6).

- Performance degradation.

- Equipment limitations.

- Transfer and spread.

Decontaminate in accordance with (IAW) the following four principles:

- Decon as soon as possible.

- Decon only what is necessary.

- Decon as far forward as possible.

- Decon by priority.

Immediate Decon

Immediate decon includes skin decon, personal wipedown, and operator's spraydown. Execute immediate decon without waiting for orders.

Skin Decon and Personal Wipedown. The individual soldier initiates decon, without command, once he becomes aware that he is contaminated.

- For CB agents, use M291 Skin Decon Kit (SDK) to decontaminate any exposed skin. Next, decontaminate your mask, hood, gloves, weapon, and individual equipment using either the M291 SDK, the M280 Decon Kit Individual Equipment (DKIE), or an M295 Individual Equipment Decon Kit (IEDK).

- For nuclear contamination, the soldier washes himself and his individual equipment, preferably with soapy water or he brushes himself and his equipment off as best he can. The primary concern is fallout in the form of dust particles.

Operator Spraydown. Begin spraydown right after personal wipedown. The spraydown removes or neutralizes contamination on the surface of equipment that you must frequently touch to perform your mission.

- For CB agents, use the M11 or M13 Decon Apparatus. Decontaminate corrosion-sensitive surfaces (radio hand microphones, precise lightweight Global Positioning System receivers [PLGRs]) with an M291 SDK, M280 DKIE, or M295 IEDK.

- For nuclear contamination, the soldier washes, brushes, or scrapes clean his equipment.

FM 31-23 (ID)

OPERATIONAL DECON

Operational decon includes MOPP gear exchange and vehicle washdown. Operational decon allows the unit to continue its mission while contaminated. It limits the transfer hazard by removing most of the gross contamination on equipment and nearly all the contamination on soldiers. These techniques do not guarantee conditions to safely allow unmasking on or near the contaminated equipment. The focus is mission accomplishment in a contaminated environment. Each battalion can conduct its own operational decon using its organic M17 Lightweight Decon System (LDS) and personnel or it can coordinate for support from the SF Group N uclear, Biological, and Chemical Center (NBCC). The detachment and FOB personnel must preplan this decon requirement before infiltration. Normally the FOB cannot support decon operations beyond a single operational decon site without augmentation.

MOPP Gear Exchange. The contaminated unit conducts its own MOPP gear exchange. MOPP gear exchange is a Skill Level 1 Common Task. Doctrinally, MOPP gear exchange is done adjacent to the vehicle washdown site. The exchange, however, could take place anywhere that mission dictates. MOPP gear exchange is the most important part of operational decon. It should take place within the contaminated life span of the overgarment.

Vehicle Washdown. Mission permitting, it is more effective to wash down the vehicles between one to six hours after contamination. There are several washdown methods available for the conduct of operational decon by the FOB or Special Forces operational base (SFOB) to enable the affected Special Forces operational detachment Alpha (SFODA) to continue its mission.

The decon team must know the contaminated unit's status—

- The number and type of vehicles.
- The number of personnel.
- The far and near recognition signals.
- The frequencies.
- The linkup point.
- The number of contaminated casualties.
- The type of chemical, if known.
- The number of replacement MOPP suits the contaminated unit has/requires.

The contaminated unit must know—

- The time required for the decon.
- The number of soldiers required to assist with the decon.

In preparation for decon, an FOB may have an element on stand-by to escort its decon team from the NBC Detachment to the decon site. The contaminated SFODA chooses the decon site. The escort element provides security for the decon team during movement to and while setting up and running the decon site. Once the site is set up, the escort element and the contaminated SFODA link up and the contaminated SFODA processes through the decon site. Simultaneous to the decon team's spraying down the vehicles, the contaminated unit conducts a MOPP gear exchange. The escort element provides security for the decon team while it decontaminates itself and closes down the site. The contaminated unit continues its mission and the escort element and the decon team exfiltrate.

When an SFODA is contaminated, the FOB may choose to infiltrate one or two members of the NBC Detachment. It infiltrates with appropriate decontaminants to decontaminate the SFODA. If using this method, the SFODA must set up and secure the drop zone (DZ) and/or LZ and pick up the decon personnel. A bundle containing NBC supplies will also be dropped along with the personnel. After decontaminating the SFODA, the decon personnel remain with the SFODA until it exfiltrates.

The decon team will infiltrate to the linkup point. Once linkup is complete, the contaminated unit will provide security while the decon team sets up and runs the decon site. The contaminated unit will drive its vehicles through the vehicle washdown. The contaminated unit will provide security for the decon team while it decontaminates itself and close down the site. The contaminated unit continues its mission and the decon team exfiltrates.

THOROUGH DECON

THOROUGH DECON includes detailed troop and equipment decon. This type of decontamination reduces contamination to negligible risk levels. It restores combat power by removing nearly all contamination from unit and individual equipment so troops can operate equipment safely for extended periods at reduced MOPP levels. Thorough decon is usually done with reconstitution and occurs in the rear area. Thorough decon is too resource intensive and time-consuming to be accomplished below FOB or Group level.

NONSTANDARD OPERATIONAL DECON

These methods are designed to provide relief from MOPP 4. They are not doctrine but tactics, techniques, and procedures (TTP) designed for a mounted mission. These methods are most valuable to the mounted detachment conducting deep operations who finds itself contaminated with virtually no possibility of an assisted decon. The detachment must rely solely upon itself for decon.

- **For Chemical Agents.** Use a five-percent bleach solution to spray down equipment that is metal or resistant to corrosion. Spray the solution only on surfaces you must touch to do your mission. Decontaminate only what is necessary. Scrub the solution into the surfaces with brushes, if available. Ideally, wait 15 minutes before rinsing off the solution. However, if mission dictates, wash off can begin with a minimum 5-minute wait time. Corrosion sensitive surfaces (radio hand microphones, PLGRs) should be decontaminated with an M291 SDK, M280 DKIE, or M295 IEDK.

- **For Biological Agents.** Scrub and rinse contaminated areas with bleach (preferred) or hot soapy water.

- **For Nuclear.** Brush or wash off dust particles from the vehicle and equipment.

FM 31-23 (ID)

DECONTAMINANT OPTIONS

Listed below are decontaminants that an SFODA can take with them on a mission and use to support nonstandard operational decontamination. Specific decontaminants for less prevalent chemical agents can be found in FM 3-9.

Uses:	A 10% solution is effective against H and VX agents, Lewisite, and biological material. A slurry mix is effective against G nerve agents.
Mix ratio:	For a 10% solution, use a mix ratio of 1-lb HTH granules to 1 gal water. For slurry, use a mix ratio of 8.5-lb HTH granules to 1 gal water.
Contact times:	Blister: 5 minutes (min). Lewisite:5 min. VX: 5 min.
Special considerations:	Must be stirred when mixed and before application to avoid settling of the mixture. Ignites spontaneously on contact with decontamination solution 2 (DS2), oils, and grease, if undiluted. Burns on contact with DS2 and VX and HD series agents, if undiluted. Corrosive to metal. Rinse with water after waiting the contact time.
Sources:	Swimming pool supplies stores HTH and commercial laundries for Calcium Hypochlorite.

Figure 6-1. High Test Hyphochlorite (HTH)/Calcium Hypochlorite

Uses:	Undiluted, it is effective against all blister and V nerve agents and biological materials.
Mix ratio:	CM agents-none, BIO-2 parts bleach to 10 parts water.
Contact times:	V nerve: 5 min. Blister: 5 min. Biological: 15 min.
Special considerations:	Burns on contact with mustard agents. Can be purchased in commercial (10-14%) or household (3-6%) concentrations.
Sources:	Any store that sells cleaning supplies.

Figure 6-2. Household/Commercial Bleach (Sodium Hypochlorite).

Uses:	DS2 in its undiluted form is effective against all known toxic chemical agents and biological material (not bacterial spores).
Mix ratio:	None.
Contact times:	All agents: 30 min.
Special considerations:	Spontaneously ignites on contact with supertropical bleach (STB) and HTH. Highly flammable 160 degrees F flashpoint, corrosive to metals, requires protective clothing when being used.
Sources:	Army supply system.

Figure 6-3. Decontamination Solution 2 (DS2).

Uses:	Effective against V and G agents, Lewisite, liquid H, and biological materials.
Mix ratio:	Slurry: 8.5-lb STB to 1 gal water or one 50-lb drum STB with 6-gal water. The STB must be added to the water to prevent boiling and splashing on the person mixing. Dry mix: 2 parts STB to 3 parts of earth or sand.
Contact times:	30 min for all agents.
Special considerations:	Mix STB only with water, stir constantly, avoid contact with skin. In dry state, spontaneously ignites on contact with DS2 and liquid blister agents.
Sources:	Army supply system.

Figure 6-4. Supertropical Bleach (STB)

Chapter 7
Mounted SFLE Operations

In Operations Desert Shield and Desert Storm, more than 800,000 military personnel from 36 nations combined their will, forces, and resources to oppose the Iraqi military. This operation, like many before and after, demonstrated the advantage of successful multinational warfare over the unilateral efforts of a single nation. The coalition increased the size of the overall force, shared the cost of waging the war among the nations, and enhanced the legitimacy of the strategic aims. In the words of General Schwarzkopf, "SF teams were...the glue that held the coalition together."

ORGANIZATION

Mission analysis will identify the number of personnel needed to conduct the mission. Joint, conventional forces, and other Army attachments may be assigned. An SFLE is an SF or joint SO element that conducts liaison between U.S. conventional forces division-level headquarters and subordinate host nation or multinational forces brigades and battalions. SFLEs do not provide combat service support to coalition forces. The parent unit of the SFLE provides logistics and administrative support.

Mission analysis also determines the number and types of vehicles required to conduct the mission. Some factors in choosing the types of vehicles are number of personnel to be transported, amount of equipment and duration of the mission, and weapons platforms.

An important consideration for a mounted detachment is contingency operations. In addition to being a liaison element, the detachment personnel may find themselves responsible for requesting, coordinating, and controlling CAS missions in the host nation unit's AO (see Figure 7-1, page 7-2). They may also be given a complete change of mission that requires them to move part or all of the detachment into another role.

The detachment should never rely solely on host nation transport. Organic vehicles should always be the first choice of the detachment. Consider rental vehicles as a way to cut the deployment costs of the operation or to reduce the signature made by American military equipment. Other nations' military units working in the AO may be able to provide transportation assistance. Host nation assets may be used for transportation needs. There can be problems—language barrier, reliability, cost, and safety—when using or relying on host nation transportation.

EQUIPMENT AND PERSONNEL PREPARATION

The following paragraphs address load plans, vehicle maintenance, and special equipment.

Vehicle Load Plan. There is no one-load plan to fit every SFLE-type mission due to the variety of mission profiles and types of vehicles associated with SFLE missions. Adhere to the same guidelines as in pre-mission planning for any mounted mission. Load plans will differ in the amount of fuel and ammunition carried. Because of the SFLE's location with or near a large friendly or host nation unit, the detachment will not have to be self-sustaining for more than a few days. Five days of self-sustainment will generally work to cover normal operations, missed resupply, and contingency operations. SFLE missions are longer in duration than mounted combat operations. Make allowances for extra equipment such as heavy equipment for base camp operations and extra personal equipment.

Monitor weight limitations, do not overload the vehicle. Ensure each individual has a seat in which to ride. If the mission requires the establishment of a base camp, avoid accruing extra equipment while in the camp that you cannot remove by organic transportation if forced to evacuate.

Figure 7-1. SFLE calls CAS in Kuwait.

Vehicle Maintenance and POL. As in any mounted operation, success or failure may depend on the detachment's ability to maintain its vehicles. SFLE vehicles typically experience less wear and tear than on a combat operation. The SFLE uses them only to move from unit to unit or base to base. Generally, the SFLE is located with or near large units from which the detachment coordinates for or uses to help maintain their vehicles. The detachment takes its normal supply of spare parts but in much smaller quantities than in a combat operation. For these reasons, fuel consumption is much less and fuel is more readily accessible.

FM 31-23 (ID)

Chapter 8
Navigational Techniques

Navigation in desert regions is more similar to navigation at sea than in other land environments. Some of the problems associated with vehicular navigation are lack of identifiable terrain features to use as reference points, outdated maps, and difficulty in keeping a vehicle on any set bearing. To minimize these problems, the mounted detachments must be thoroughly versed in the four levels of mounted navigation, each level supplementing the other. These four levels of navigation are vehicle orienteering, DR, celestial position fixes, and satellite position fixes. See FM 21-26, Map Reading and Land Navigation, 7 May 1993, chapters 6 and 12.

NAVIGATOR'S DUTIES

The mounted detachment uses one primary navigator who is located in the lead vehicle. He is usually the most experienced vehicle navigator and route planner. His primary duty is to ensure the detachment arrives at the appropriate destination(s) at the right time(s). He accomplishes this duty by completing numerous subtasks, such as—

- Planning the route(s) to use with the detachment commander. This planning includes tentative laager sites.

- Keeping a log in which he records planned and actual time, distance, and direction. He can plot or chart this data at convenient intervals to ensure correct course and to estimate times and duration(s) for future movements.

- Estimating, on short notice, the detachment's estimated position within a reasonable degree of accuracy (400 meters using DR, 200 meters when vehicle orienteering, or 100 meters using satellite position fixes).

- Making frequent checks on his estimated position using satellites, bearing fixes, or celestial fixes.

- Finding his objective by methodical search if it is not located when reaching his estimated position to the objective.

The navigator uses general and specific maps. General maps are for route planning, general navigation, and plotting fixes. General maps usually used are Joint Operations Graphics at 1:250,000 scale, and Geological Survey maps at 1:100,000 scale. Specific maps are Defense Mapping Agency (DMA) at 1:50,000 scale and United States Geological Service (USGS) at 1:24,000 or 1:62,500 scale. The navigator uses sterile maps for operational security. Map sets required for a mounted operation are considerably larger than those used in a standard dismounted mission. Listed below are some tips for working maps.

- Cut off all unneeded map borders to decrease the map's size for use inside the vehicle.

- Use combat acetate to protect both sides of the map sheet(s). This acetate increases the life of the map and allows the navigator to mark the map using alcohol pens, grease pencils, or other tools that can be erased easily without destroying the map.

- Use a map storage container to maintain positive control of the map set and to prevent and limit damage to the map(s). This container can be a map book made out of meals, ready-to-eat (MREs) box sides with the maps attached to the book's "pages" or a polyvinyl chloride (PVC) pipe strapped to the ceiling of the vehicle with opening toward the driver and the navigator to store the map sheets.

- Store all the tools (pencils, grease pencils, alcohol pens, and protractors) within easy reach of the navigator working inside the vehicle.

The primary tools the navigator uses, other than the maps, are the vehicle compass, odometer, and GPS. He can also use a sextant or like tool for celestial navigation to support his other tools. He must be proficient with all of these devices. He cannot depend on one device alone; the tool he is counting on the most will be the one to break when it is most needed.

TERRAIN ASSOCIATION

When the detachment moves though terrain with readily identifiable terrain features, terrain association is the preferred method of navigation. The primary navigator plans his route so that the detachment moves from terrain feature to terrain feature.

Consider the tactical situation. Select concealed routes to avoid skylining.

Consider ease of movement. Use the easiest possible route and bypass difficult terrain. A difficult route will be harder to follow, be noisier, cause more wear and tear (and possible recovery problems), and take more time. Try to select a corridor instead of a specific route. Make sure the detachment has enough maneuver room.

Use terrain features as checkpoints. These must be easily recognizable in the light and weather conditions and at the speed at which the detachment moves. Find a terrain feature that can be recognized from almost anywhere and used as a guide.

- The best checkpoints are linear features that cross your route. Use wadis, rivers, hardtop roads, ridges, valleys, and railroads.

- The next best checkpoints are elevation changes, such as hills, depressions, spurs, and draws. Look for two contour lines of change. You will not be able to spot less than two lines of change while mounted.

- In wooded terrain, try to locate checkpoints at no more than 1,000-meter intervals. In open terrain, you may go to about 5,000 meters.

Determine Directions. Break the route down into smaller segments and determine the rough directions to follow. You do not need to use the compass; just use the main points of direction (north, northeast, east, and so forth). Before moving, note the location of the North Star (or Southern Cross, if below the equator). Locate changes of direction, if any, at the checkpoints picked.

Determine Distance. Get the total distance to be traveled and the approximate distance between checkpoints. The navigator uses the speed and time method and the odometer count to determine distance traveled.

- **Speed and time method.** This method is the least desirable because of the need to keep very accurate records of vehicle speed. The navigator computes distance traveled by multiplying the constant vehicle speed by the hours and tenths of hours spent traveling to get total distance traveled.

	------	constant vehicle speed
multiplied by	------	hours/tenths of hours traveled
equals	------	total mi traveled

- **Odometer count.** The preferred method for measuring distance. Before the detachment can rely on the odometer, it must be tested at a known distance of at least two miles. Accuracy should be exact on hard surface roads. Soft sand or loose rocks will cause what is called "wheel slip." Wheel slip is when the vehicle's wheels turn in overproportion, causing the odometer to read greater distance traveled than the actual distance traveled. Wheel slip factor comes with experience, but a general rule is that moderately soft sand will cause the wheel to slip up to 10 percent. Upon determining the wheel slip factor, the navigator multiplies it by the distance to be traveled. The result obtained gives him the odometer reading when the detachment arrives at the destination.

	------	distance to be traveled (statute mi)
multiplied by	------	wheel slip factor
equals	------	odometer reading when reaching destination

If the navigator can determine distance traveled, he then needs a method for keeping the vehicle on a bearing (azimuth). The navigator has three primary tools at his disposal to maintain azimuth:

- The liquid-filled, vehicle-mounted compass (adjusted to account for the vehicle's electrical field while engine is running).

- The satellite positioning device.

- The individual soldier's lensatic compass. (This compass can be used inside the vehicle if the user accounts for the amount of declination caused by the vehicle and the compass is used in the same position on the vehicle every time. The electrical field in a running vehicle can throw off a compass 25 to 30 degrees and it is different in every part of the vehicle.)

After determining the correct azimuth, the navigator orients the driver to the direction of travel. The navigator does this by picking a point in the distance and identifying it to the driver. This point can be a terrain feature, a man-made object, or a celestial object.

Make Notes. Mental notes are usually adequate. Try to imagine what the route will be like and remember it.

Plan to Avoid Errors. Restudy the route selected. Try to determine where errors are most apt to occur and how to avoid any trouble.

Use a Logbook. Another tool is the navigator's log, also called the driver's log. It lists checkpoints and distance traveled and to be traveled. It can also list azimuths or direction in cardinal points of magnetic degrees. This log can also list wheel slip factor. The navigator's log in Figure 8-1, page 8-4, lists checkpoints by their numbers (memorized by the navigator), their location on the map, and distance to be traveled to the next checkpoint. Only the navigator needs to memorize the checkpoints. He can refer to the log for instructions.

CHECKPOINT	CHECKPOINT	DISTANCE	NOTES
49 Columbus	50 Ranch	17.0	
50	51 Johnson Tank	10.2	
51	52	2.2	track W
52	53	3.9	track N
53	54	2.8	track N
54	SS Landing Strip	6.7	track N

Figure 8-1. Example of a navigator's log.

Another important issue to remember is the navigator must update the driver and gunner to the direction of travel, distance of travel, rally points, and checkpoints along the route. This updating must be done in case a situation arises demanding immediate action so that the entire crew will know what to do and where to go. Should the detachment split (break contact) and the navigator is injured, the remaining vehicle crews must know their location to conduct a linkup with the other detachment members. The linkup plan must be planned and rehearsed in isolation.

DEAD RECKONING

DR is moving a set distance along a set line. It is the general navigation technique used when there are no terrain features on which to take bearing fixes or when the region in which you are traveling is uncharted or poorly mapped. The detachment will normally use a combination of terrain association and DR to navigate.

When using DR, use a navigator's log to ensure ease of transition from different bearing land distances. The navigator relies solely on direction, usually in magnetic degrees and distance traveled, to plot his position from the known starting point (SP).

While using DR, it is essential that the navigator maintain an accurate account of distance traveled.

VEHICLE ORIENTEERING

Vehicle orienteering over unimproved road networks or cross-country consists of terrain association and DR, bearing fixes, and use of the navigator's log. Remember that there may be very few prominent terrain features in some areas where you may have to navigate. Therefore, the only way you can confidently navigate is to use all these techniques together. At this point, also remember to use your gunner as a navigational tool. From his vantage point in the turret he can see terrain or man-made objects sometimes not visible from the navigator's point of view.

If at any time during movement you locate a readily identifiable terrain feature shown on the map, take a bearing fix. This fix is to check or correct the position of the detachment. These bearing fixes can be single, multiple, or running.

Single bearing fix (modified resection). This fix will determine that your approximate DR position is somewhere on the line of bearing.

FM 31-23 (ID)

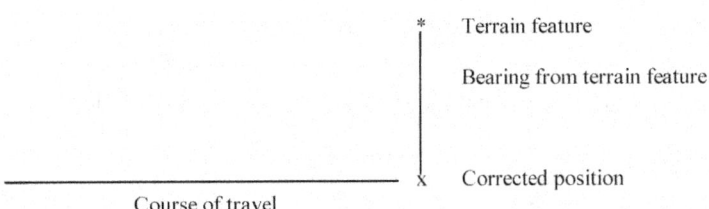

- Orient the map using a compass or by terrain association or DR.
- Find a distant point that can be identified on the ground and on the map.
- Determine the bearing (magnetic azimuth) from your location to the distant known point.
- Convert the magnetic azimuth to a grid azimuth.
- Convert the grid azimuth to a back azimuth. Using a protractor, draw a line for the back azimuth on the map from the known position back toward your unknown position.
- The location of the user is where the line crosses the detachment's course of travel.

Multiple bearing fix (resection). This fix will produce two or more bearings that will intersect, showing the exact corrected position from where the bearings were taken. Resection is the method of locating the detachment's position on a map by determining the grid azimuth to at least two well-defined locations that can be pinpointed on the map. For greater accuracy, the desired method of resection would be to use three or more well-defined locations.

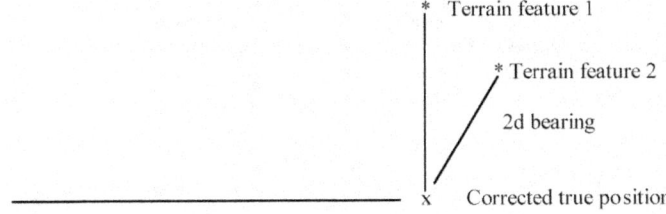

- Orient the map using the compass.
- Identify two or three known distant locations on the ground and mark them on the map.
- Measure the bearing to one of the known positions from your location using a compass.
- Convert the magnetic azimuth to a grid azimuth.
- Convert the grid azimuth to a back azimuth. Draw a line for the back azimuth on the map from the known position back toward your unknown position.

8-5

- Repeat for a second position and a third position, if desired.

- The intersection of the lines is your location. Determine the grid coordinates to the desired accuracy.

Running bearing fix (combination intersection/resection) This method is performed during movement when few prominent identifiable terrain features are visible. A running bearing fix will determine the detachment's approximate DR position with a minimum delay in movement.

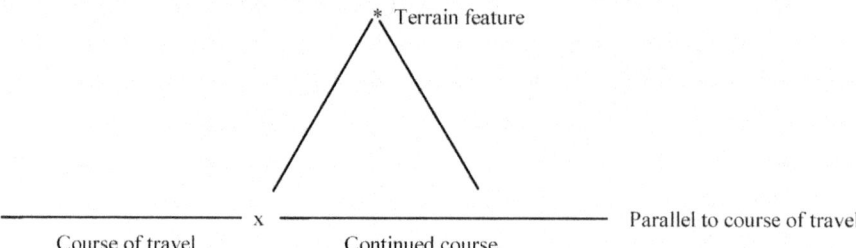

- Identify a terrain feature located on the map.

- Take a bearing to the terrain feature.

- Convert the magnetic azimuth to a grid azimuth.

- Convert the grid azimuth to a back azimuth. Draw a line for the back azimuth on the map from the known position back toward your unknown position that crosses the detachment's route of travel.

- Continue on route azimuth to a position where the terrain feature is again in sight that is 30 or more degrees from the first location. During this movement, maintain an accurate measurement of distance traveled.

- Take a second bearing to the terrain feature.

- Repeat the steps above to plot the bearing on the map.

- Using the distance traveled from the 1st and 2d bearing fixes, create a scale that represents this distance on the map.

- Orient the scale on the map so that it is parallel to the detachment's course of route on the map.

- Move the scale up or down until it intersects the 1st and 2d bearing fix. Where the scale intersects the second bearing fix is the detachment's location.

Methodical Search. Use this method when the detachment has traveled its planned distance and the objective is not readily apparent. Usually, when this occurs, the detachment has not traveled far enough, wheel slip is greater than anticipated, or the distance was computed short.

The detachment stops upon traveling the planned distance. When it does not discover its objective, it implements the square search by first determining the visibility. It then travels on the same bearing for the distance of visibility. It then makes a right or left 90-degree turn and travels perpendicular to the original bearing for a distance twice that of the visibility distance. It keeps making right or left 90 degree turns, traveling three, then four, then five times the distance of visibility until it spots the objective.

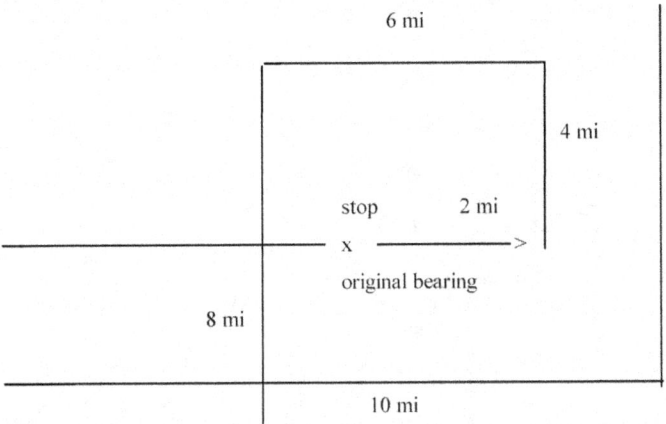

Square search depicted, visibility is two miles. Detachment conducts an increasing square shap search pattern to methodically locate the objective.

CELESTIAL NAVIGATION

Celestial navigation is another navigation tool available to the mounted detachment. It requires the greatest degree of training to use and to maintain proficiency. In simple terms, celestial navigation is taking altitude readings from celestial bodies (stars) and computing these readings based on time of reading (observation) to determine a line or lines of position that can be plotted on a map to triangulate or intersect your position. It is not a fluid form of navigation but rather a means of fixing or determining your position after stopping. There are two types of celestial fixes—single position line (line of latitude) and multiple position line (lines of intercept).

Single position line or observation for latitude. These are altitude readings and computation of Polaris (North Star), the Sun, or the Moon. From an observation of one of these celestial bodies, we can determine the latitude meridian that we are on. This observation is of particular value when we are positioned on a known north-south line.

Multiple position line or lines of intercept. From altitude readings and computations of three of the fifty-eight accepted stars, we can determine and plot intercept lines. From these plots, we can determine our triangulated position. The mounted detachment, moving mostly at night, will stop before sunrise and observe three star shots to determine position. From this stop, the detachment can move a short distance away keeping exact distance and direction to plot the laager site from the triangulated position. The mounted navigator then plots his position using the three fixes.

For observation and computations of altitudes of celestial bodies, the detachment uses—

- A small portable sextant to make altitude readings or a theodolite to make altitude and bearing readings.

- A nautical almanac for year in use to make computations for the celestial bodies.

- Sight reduction tables for the latitude range operating in. These are also used for computations.

- Computation forms for computations of the moon, the sun, and stars.

- Plotting instruments, a protractor, triangle, and parallel ruler to plot intercept lines.

SATELLITE NAVIGATION

Satellite navigation is the most popular and easiest method of navigation used by mounted detachments. A GPS is the most accurate means available to determine your location on the battlefield at all times. Using a GPS allows you many options by which to navigate.

The GPS can show magnetic azimuth, continuous position fix, and vehicle speed to aid in navigation. A navigator only relies on the GPS to back up his base navigational skills. Using his map, compass, and odometer readings, along with terrain orientation or DR, the navigator uses the GPS to confirm or make corrections in his route movement when needed. This method is the preferred method of use; do not rely solely on the GPS. A problem with the GPS such as power outage, a broken antenna, or the loss of satellite reception could leave you disoriented. You should be able to successfully navigate to your objective using the techniques previously addressed in this chapter and combining them with the GPS.

Another use for the GPS is to store waypoints. A waypoint is the coordinates of a specific location in your route programmed into the GPS. Once there are two or more waypoints, the navigator can set the GPS to plot a route from a given point to another given point. When done, the GPS gives direction in degrees magnetic, distance to travel, and the time it will take at the vehicles' current speed to arrive at the desired location. The device also indicates when the vehicle is off course due to wheel slip and allows the navigator to make a correction. He can tell the driver how much of a correction to make. Before departing isolation, the navigator can preprogram the detachment's entire route into the GPS for navigational purposes. During movement the detachment commander can designate a location as a rally point, water source, target reference point (TRP), or another point of interest. The navigator can store this particular location as a waypoint. This storage serves a dual purpose because you can retrieve this waypoint if you need to navigate back to this location or just need the coordinates for operational use.

The GPS provides you with the ability to rapidly obtain an accurate polar plot to a target from your position.

After learning to use the GPS and seeing what a powerful tool it can be, do not become solely dependent on the device. Its use can be lost at any time due to mechanical problems. Using the GPS as an aid to check your position should be the preferred method of choice, but always depend on the basics and you cannot go wrong.

Any GPS is subject to command navigation warfare. It is possible for the enemy to produce false signals that will cause your GPS to not work or produce inaccurate information. This is an easy problem to correct, but the navigator must be aware of the possibilities. If you suspect that you are a target for this kind of information warfare, dig a hole below ground level and place your GPS antenna into the hole to check your position. This hole must be deep enough to block any line-of-sight ground based transmissions. The antenna will only receive signals from satellites overhead and will give you a correct navigational reading.

We do not speak on the operation of a particular model of GPS because there are many types in use today. Therefore, take it upon yourself as a detachment member to train up on the particular equipment organic to your unit or get training on equipment you might receive before a mission.

FM 31-23 (ID)

Chapter 9
Camouflage

SF mounted detachments operating behind enemy lines will have to stay undetected to complete the mission. In an unsupported role in a desert environment, the only way to remain undetected is using proper camouflage measures. Proper camouflage is critical for the detachment operating behind enemy lines with no support or limited outside support. The detachment's ability to hide in the desert is limited only by the imagination and resourcefulness of its members (see Figure 9-1).

Figure 9-1. Camouflaged GMVs.

CAMOUFLAGE THEORY

The biggest threat to the detachment is detection. Detection can be by—

- **Direct observation.** Where the observer sees the subject with his eyes, either aided or unaided.
- **Indirect observation.** Where the observer sees an image of the subject and not the subject itself. Indirect observation uses photography, radar, infrared, thermal imaging, and tele-video.

Regardless of the method of observation, certain factors help the eye and brain identify an object. The six factors of recognition are—

- **Position.** This factor relates to the position of the object in relation to its surroundings. In addition, position is space relative to one object and another.

- **Shape.** Experience teaches people to associate an object with its shape or outline. At a distance, the outline of objects can be recognized long before the details of its makeup can be determined. Trucks, guns, tanks, and other common military items all have distinctive outlines that help to identify them.

- **Shadow.** Shadow may be even more revealing than the object itself. This fact is true when viewed from the air. Sometimes it may be more important to break up or disrupt the shadow than the object itself.

- **Texture.** Texture refers to the ability of an object to reflect, absorb, and diffuse light. It may be defined as the relative smoothness or roughness of a surface. A rough surface reflects little light and will usually appear dark to the eye or in a photo. A smooth surface such as an airstrip, although it might be painted the same color as its surroundings, would show up as a lighter tone on a photo. One of the most revealing breaches of camouflage discipline is shine. Shine attracts attention by reflecting light such as sunlight or moonlight.

- **Contrast.** Color is an aid to an observer when there is a contrast between the object and its background. The greater the contrast in color, the more visible the object is. Usually darker shades of a given color will be less likely to attract an observer's attention than the lighter shades.

- **Movement.** The last factor of recognition is movement. Although this factor seldom reveals the identity of an object, it is the most important one of revealing location. Movement is detected easily and usually through the observer's peripheral vision.

CAMOUFLAGE METHODS - CONCEALING OBJECTS

Hiding is the concealment of an object by some form of physical screen. Hiding is accomplished by using thick vegetation or terrain features that screen vehicles from ground observation. In some cases, the screen itself can be invisible to detection and, at times, it is the overt screen that protects the activity or equipment from observation.

Blending is the arrangement or application of camouflage materials on, over, or around an object so that it appears to be part of the background. Blending distinctly man-made objects into a natural terrain pattern is necessary to maintain a normal and natural appearance.

Disguising involves the simulation of an object or activity so that it looks like something else. Clever disguises will mislead the enemy as to identity, strength, and intention.

CAMOUFLAGE IN THE DESERT

There are camouflage problems encountered in the desert that require special attention to overcome. The lack of natural overhead cover, the increased range of vision, and the bright tones of terrain all require emphasis on siting, dispersion, and camouflage discipline to achieve concealment. Cast shadows are notably conspicuous.

Deserts the world over have, in general, extensive areas of sand, lack of tall vegetation, brilliant sunlight, and extreme temperature ranges. Rocky areas, steep wadis, and washes are all characteristics of desert environments. The density of vegetation coverage is often as high as 80 percent. Most of the vegetation is low, averaging about 30 inches high in flat areas, while in the wadis and at higher elevations, it can average close to 10 feet. When viewed from the air, the desert floor appears spotted or pockmarked in many areas.

Vegetation commonly found in the desert includes colors ranging from pale yellow to dark gray and dark brown. Although green and brown are the principal colors of most desert vegetation, it is important to study the target area vegetation and terrain to formulate a proper vehicle camouflage plan.

No one camouflage system or pattern will work for every desert or even different parts of the same desert. Only with detailed planning can a mounted detachment plan for and prepare the materials necessary to properly conceal their vehicles.

CAMOUFLAGE CONSIDERATIONS

In preparing for desert operations, position selection, reflection reduction, and concealment are conditions the detachment must consider.

Position Selection

Siting or position selection is of critical importance in any environment but particularly so in the desert. Site positions that fit into the existing ground pattern with minimum alteration to the terrain. The sites selected should suppress ground observation. Some areas such as valley floors might have sparse vegetation, but adjacent wadis could offer thicker vegetation with opportunities for defilade and enhanced potential for concealment from aerial threats. Day laagers should not be areas that would be obvious to enemy patrols. The operations sergeant usually positions the vehicles to provide 360-degree security, good concealment, and to allow rapid egress from the position.

Reflection Reduction

Reducing surfaces that reflect light is a measure that starts in garrison before deploying by removing mirrors and covering headlights and taillights. Normally the windshield is not removed so that it can provide protection from blowing sand, dust, and rocks thrown by the vehicle in front. Detachment members cover all reflective surfaces with a close weave, non-see-through cloth (canvas or target cloth). A sight portal must remain open for driving. If cloth or other material is not available, mix water and dirt to get mud and apply it to the reflective surfaces.

Concealment

Usually the most effective way to conceal vehicles is by the use of netting. The Light Weight Camouflage Screening System (LWCSS) is preferred in the desert. This net provides concealment from visual, near IR, and radar and target acquisition devices. This net is not intended as a complete camouflage system as it depends on its imitation of the ground surface, both color and texture, to be effective. In some deserts, the woodland pattern would offer greater ability to blend in. Alternatives to the LWCSS are—

- Open weave cloth with patchwork colored to match the terrain in the operational area. This type of net might be the preferred choice if operating in a predominantly sand dune area.

- Large fishing net garnished with burlap to suit the color of the operational area. Vegetation can be added to this net to enhance concealment.

NOTE: When using netting in open areas, drape the net over the vehicle and slope the sides gradually to the ground. Break up the outline of the vehicle by placing props or poles underneath and intertwine vegetation into the net. Eliminate shadows caused by the vehicle or net.

In broken country, use the drape to tie the net to some irregularity in the terrain, such as next to a mesquite or brush mound. Break up the outline and eliminate shadows.

After placing the net, cut and place brush into the net to add realism, texture, and similarity to the terrain and to help break up the outline.

Chapter 10
Maintenance and Recovery

Maintenance is the most important support service in mounted operations. Long supply lines and minimum stocks on hand will increase the time needed to get vital replacement items and repair parts. It is imperative that proper maintenance be performed on equipment throughout the whole spectrum of service (before, during, and after operations). See Figure 10-1.

Figure 10-1. Vehicles on line in 5th SFG(A) motor pool.

GENERAL

The maintenance organization functions essentially the same as in other operations; however, the effects of the hot climate and the abrasive, windblown sand on equipment will increase all maintenance requirements.

The mounted detachment should prepare itself to handle all maintenance required at operator and organization level. In addition, some depot level knowledge is necessary. Each new member should attend a maintenance course for the GMV and DOM.

Team personnel must receive training at the unit motor pool under the tutelage of the battalion maintenance section. It is incumbent on the detachment personnel to become mechanics for all their own equipment. The detachment leadership should become familiar with The Army Maintenance Management System through self-study, on-the-job training, or correspondence courses.

Designated motorcycle riders must receive proper maintenance training for their motorcycles. The detachment cannot expect to find maintenance facilities inside the operational area.

PREVENTIVE MAINTENANCE CHECKS AND SERVICES

The vehicles assigned to a mounted detachment are its most important assets. Its members must perform routine PMCS on their vehicles before, during, and after all operations.

The vehicles also require exercise. If a team deploys without its vehicles or it spends an extended period at home station without using their vehicles, it must arrange for PMCS, and for starting and exercising its vehicles.

The team must perform post-operations maintenance procedures immediately after the conclusion of its mission (see Appendix L).

DESERT ENVIRONMENTAL EFFECTS

Several factors will affect mounted operations in a desert environment. The following paragraphs address these factors.

Rough Terrain

Severe terrain consisting of rough, uneven ground, steep mountains, and loose sand and rocks will cause vibrations and result in the loosening of nuts and bolts and fuel and hydraulic lines. It could also disrupt electrical components. Rough terrain can severely affect wheels, transmissions, and suspension systems. Therefore, frequent inspections and maintenance periods are necessary to ensure vehicles function properly and to prevent long downtime due to repairs.

Sand and Dust

The abrasive effects of sand and dust adversely affect equipment. Any moving part faces the probability of being damaged or impaired by sand or dust. Brakes, recoil systems, bearings, hydraulics, and relays are all susceptible to incapacitation by sand or dust. Also sand and dust mixed with lubricants turns into an abrasive paste that can easily wear and score moving parts. Cover equipment when not in use. Frequent preventive maintenance will help to alleviate these problems to a manageable degree.

Heat and Low Humidity

Intense heat and low humidity can cause overheating of the vehicles and batteries and the degradation of seals and tires. Surface temperatures heat parts and accessories making them untouchable without protection. Surface temperatures can reach 140 degrees and reflect heat under and into vehicles. Again, frequent inspections, protection with covers, and regular maintenance will reduce the effects of these environmental factors.

Vegetation

In some deserts, thorny and spiny plants pose a problem to tires, especially at night. They can puncture radiator hoses. Unsecured or unprotected equipment can become a victim of vegetation also. Secure all equipment to the vehicles. Individual driving technique is the first preventive measure for stopping flats.

FM 31-23 (ID)

LESSONS LEARNED

The following paragraphs address different areas of the vehicles and what was learned during actual operations. The mounted detachments apply the results of these lessons learned to better prepare themselves for operations.

Filters. Clean all filters regularly to maintain engine efficiency and avoid complications. Use fuel filters or strainers when refueling to avoid fuel contamination and clogged fuel lines.

Tires. Keep the tires at proper tire pressure (20 pounds per square inch [psi] [front] and 22 psi [rear] for standard HMMWV tires) and filled with industrial sealant to avoid flats. Carry extra tire plugs and repair kits.

Generators. Inspect generators daily for wear and loosening of the shafts. If the pulley or shaft breaks, it is driven downward and results in a ruptured cover for the steering gear box and loss of power steering fluid. If the generator is loose and cannot be fixed correctly, then take it off. The GMV can run at least seven days, when operating at night, without a generator. Change batteries with other GMVs that have good generators.

Batteries. Check for leaks and evidence of cracks. Carry distilled water in a nonmetallic jug to replace battery fluid. In hot weather, batteries discharge more electricity, necessitating checking specific gravity. In cold weather, coat battery terminals with grease to protect against the cold.

Tie Rods. Reposition tie rod retaining clamps so that they will not rub the tire when turning the wheel. Rubbing will cause tire failure or puncture. Turn these clamps toward the inside.

Glow Plugs. Be careful when replacing glow plugs. They have a tendency to swell at the end, causing it to break off when removed. If the glow plug appears hard to remove, remove the injector directly above the plug. Coat the interior space with grease, then remove the plug. The grease keeps the broken glow plug from falling into the heads.

Radiators and Fan Belts. Overheating is a major problem in the desert. Overheating is a greater problem moving during daylight than at night. Keep the fan belts at the right tension. Keep the radiator free from debris. Use a corrosive inhibitor in the water and coolant mixture. Inspect the water pump shaft bearing often to determine if it has worn bearings.

Brakes. During travel over rough terrain, the bolts and nuts holding the brake calipers can become loose. Check these often for tightness. Vibration from braking and front-end whining will be the first indications of loose calipers during movement.

Drive Train. Keep the drive train and U-joints lubed properly and remove excess grease to prevent sand from sticking. Moving through sand and dust will make the lubing more important and frequent.

Filler Caps. Remove all sand and dust from the caps before removing them to fill fluid levels. This action will prevent contamination.

Fuel Tank. Check fuel lines, top clamps, and vent line hoses for tightness. During a long mission, loose clamps will cause fuel leaks that reduce mileage.

Turret Ring. Check if turret sticks or is hard to turn. Normally the turret ring gasket coming off track causes this. Removing this gasket usually remedies this problem; however, it must be noted and repaired as soon as possible.

OFF-ROAD DRIVING

Good off-road driving technique is the first preventive step in limiting broken vehicle parts or becoming stuck. All drivers must become well trained in judging terrain and negotiating various ground conditions. Most detachment movements will be at night, so driver's training should focus on the use of night vision devices. In addition, drivers should develop the following skills:

- Selecting proper gear ratio and shifting.
- Using momentum.
- Knowing the vehicle's capabilities.
- Estimating and using proper speeds.
- Avoiding sudden forward and braking thrusts.
- Applying traction theory.

Drivers must become familiar with varying terrain conditions found in the desert and considerations for crossing the conditions encountered.

Sand Dunes

In nonvegetated sandy areas, the wind can sweep sand, packing it into high sand dunes. These high sand dunes can be extremely high and steep and almost impossible to traverse when fully loaded. Sand dunes can form a crust on the surface, usually about two inches deep, that makes the dune surface appear to be hard. If crossing under these conditions, the surface will break under the vehicle's weight leaving the vehicle stuck in loose sand. Avoid crossing sand dunes at all. If crossing is necessary, conduct a reconnaissance, if possible, to determine the best route to limit chances of becoming stuck. The best bet is to traverse these large sand areas by driving around the sand dunes at their lowest point.

To increase traction when driving in sand, the driver can reduce tire pressure all the way around. Take care not to reduce the tire pressure too much, the tire will come off the rim. When reducing tire pressure, the tire's footprint increases, giving the vehicle more flotation or surface area to grip the sand.

Remember to re-inflate the tires to correct operating pressure once clear of sandy areas. Low tire pressure at normal operating speeds is dangerous and has been known to cause the GMV to turn over.

When approaching large areas of deep sand, increase vehicle speed before you enter and keep the vehicle's momentum steady. Do not turn the vehicle's wheels sharply as this move can cause loss of momentum.

FM 31-23 (ID)

Rocky Areas

Rocky and boulder-strewn areas may extend for miles in all directions. Rocks in desert environments are very often sharp-edged due to erosion or from volcanic origin. These sharp rocks are very hazardous to tires. Often rocks are so numerous that it is impossible to avoid all but the largest ones. Driving in such areas causes extreme wear on tires, suspension, and drive train components. The shock incurred when traversing rocky areas can also break equipment stored on the vehicle if not secured properly.

Tire pressure can be lowered to reduce the bumpy ride and shock that is transferred to the vehicle. However, lessons learned indicate that a higher than normal tire pressure helps reduce punctures from smaller rocks. Take care not to let rocks scrape the wall of the tires when driving in rocky areas, this may cause a sidewall puncture that is difficult at best to repair.

Wadis

These are dry riverbeds caused by fast moving runoff water from higher elevations after a rain. Most wadis have a smooth bed and prove to be excellent tracks for travel, but never travel in a wadi when it has been or is raining, due to flood danger. When coming across a wadi, look for a good entry and exit point. Many times the banks will be steep. If the wadi is not too steep or narrow, cross it by entering head on or at a slight angle. Be careful not to turn the wheels too much in any direction so that if the vehicle hits a hole or slides down into the wadi, the wheel will not be torqued and break the ball joint. Use a ground guide to walk the path then ease the vehicle into the wadi using low gear.

Small Ditches and Rises

Cross small ditches at an angle to prevent the vehicle from becoming high centered. Enter these obstacles at a low speed. High-speed entry may cause the vehicle to tip or roll over.

Salt Marshes

These areas are mostly impassable due to the powdery silt and wet, muddy areas. Mud-packed tire treads will deny traction. Although salt marshes should be avoided, small areas not on maps may have to be crossed. Use rocks, sandbags, perforated steel planking (PSP), or dry sand to construct a passable bed. Loss of momentum in sand results in getting stuck.

RECOVERY

All recovery during operations by the mounted detachment will consist of self-recovery methods—either when it becomes stuck or when it has a mechanical breakdown (Figure 10-2, page 10-6).

The most prevalent cause of a vehicle becoming stuck is driver error. A major cause of equipment breakdown and/or malfunction is poor maintenance, pointing to driver/crew error. It is very important to use good driving techniques and proper vehicle maintenance. When this fails, the detachment must recover the vehicle.

Vehicle recovery is easiest when the tires still have traction and it is assisted back out through the original tire tracks. Use a second vehicle to winch out the stuck vehicle. The winch has a 6,000-lb capacity. The winch is used only to assist vehicles, never as the sole source of power.

FM 31-23 (ID)

Figure 10 2. Repairing a flat tire.

The detachment should carry tow straps or chains. Braiding rope (three 12-foot by 5/16-inch pieces) or a 20-foot chain will work well. They should have hooks or clevises attached to the ends for anchoring to the vehicle. If possible, a detachment carries at least one tow bar to assist in long-range recovery or when towing a vehicle at high speed.

When a vehicle is stuck in mud or sand, use the pioneer tools to emplace dry or solid matter under the tires for traction. Sand bags or PSP can be dug into and under the wheels to assist traction. The detachment should carry empty sandbags for this purpose.

When conducting recovery, one section provides security as the other vehicle makes the recovery. Always decide beforehand where the vehicle is going after breaking it loose.

The GMV has a 6,000-lb capacity winch, with a 100-foot long, 3/8-inch cable. A remote cable operates the winch (unwinding or winding the cable on the spindle). There is a neutral power lever on the winch itself for extending the cable under power. When using the winch, remember these do's and don'ts—

- Use the vehicle's wheel power to help the winch.
- Don't overtake the cable.
- Carefully prepare the winching operation.
- Be careful of personnel positioning should the cable snap or unhook.
- Make sure the anchor points are solid.
- Don't exceed the maximum angle of pull.

- Use artificial surfaces for traction when stuck in water or soft sand.

The detachment makes contingency plans for what to do with vehicles they are unable to repair or recover. It makes every attempt to recover the vehicle and return it to a place where it can be exchanged or repaired. If unable to recover the vehicle, it will normally be destroyed in place to prevent it from being captured by the enemy.

Chapter 11
Logistics

Mounted detachments can operate for long periods without external resupply, depending on the duration and distance of the mission. The normal mission planning range for a standard GMV detachment is 10 days or 500 mi. However, with the onboard cargo capacity of the GMV and the DOT, the detachment can pack enough supplies and fuel for 10+ days or 1,000 mi.

GENERAL

The extended supply lines required for expanded distances calls for special considerations and procedures to ensure adequate supplies for the mission. Resupply is provided in one of the following ways:

- Onboard supplies.
- Caches.
- Airdrops.
- Sea resupply, if near a coast.
- Conventional unit linkup.
- Linkup with an AOB at an MSS.

Normally the AOB provides the best way to resupply the mounted detachment behind enemy lines, especially if the FOB does not want the mounted detachments to recross borders or FEBAs but must resupply the detachments for follow-on missions. The AOB performs this resupply using the MSS explained in detail below.

The mounted detachment places emphasis on fuel, water, mission-essential equipment, ammo, and demolitions. The detachment must compute and carry enough fuel, water, PLL, and ammo for mission accomplishment. Appendix F lists a generic mission profile of 10 days or 1,000 mi.

POL/PLL requirements are best determined through experience but must address parts that are broken the easiest and would deadline the vehicle. Carry enough fluids, lubricants, and fuel for mission duration. GMVs with less than 10,000 mi typically do not use oil but the detachment must carry enough oil to tend to engine use, plus enough to replace all engine oil in case of catastrophic damage to the oil pan.

MISSION SUPPORT SITE

The mounted detachment can only sustain itself for a certain period. A longer duration or a follow-on mission will require a resupply. Sometimes the most available means of resupply involves a linkup with the AOB at an MSS. This method gives greater flexibility to the mission(s) because the SFODB running the AOB has identical training, equipment, and operational procedures. There are two types of AOB MSSs:

Laager MSS (Figure 11-1). Ideally, SFODA links up with a laagered SFODB. The SFODA integrates into the perimeter. As the SFODA enters the perimeter, it is guided to a sister vehicle from the SFODB. After camouflaging, they conduct resupply and maintenance activities assisted by the SFODB. The SFODA stays at the MSS until all services are completed and then vacate under hours of darkness.

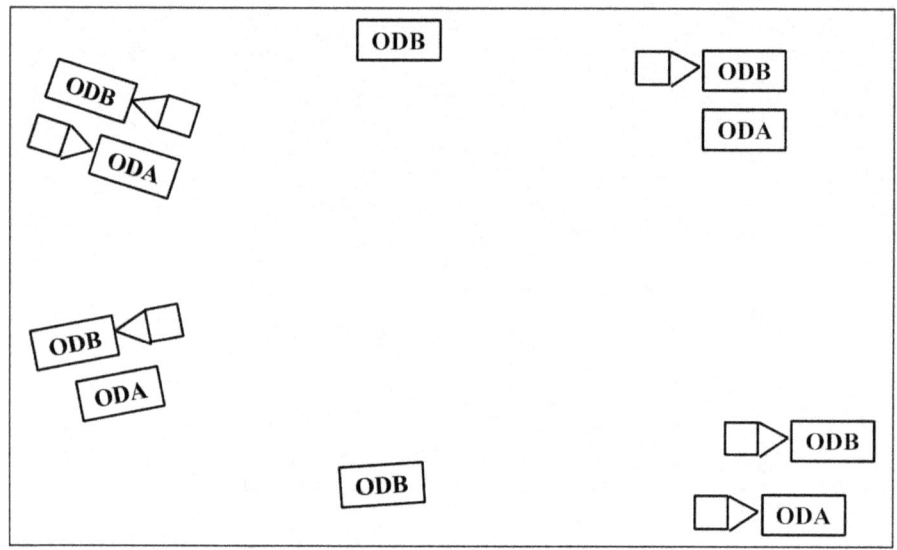

Figure 11-1. Example of a laager MSS.

Gas Station or Fluid MSS (Figure 11-2, page 11-3). This resupply MSS is arranged in a linear pattern. Individual vehicles move through and are serviced at each logistics maintenance station. They then reassemble at the exit point holding area. Choose this type of MSS in restrictive terrain and when the SFODA needs to resupply quickly without loosing security. Once reassembled, the SFODA continues on its mission and the MSS is quickly broken down and sterilized. Due to the dispersion of the gas station, extra security is necessary.

FIVE R's

Refuel. The first requirement is to refuel the SFODA. Refueling is by 5-gal cans or from bladders installed on the MSS element's vehicles. If refueling with cans, replace the empty cans on the SFODA's vehicles. If using bladders, fill the empty cans. Try to limit the time the SFODA spends in the MSS.

Rearm. The SFODA requests ammo by amount and type before the MSS is established. The MSS element will normally carry a stock of spare weapons parts and, in extreme cases, will replace broken weapons with its own.

Refit. This service includes replacement or repair of detachment equipment such as radios, medical gear, NBC equipment, or other items based on a precoordinated stockage determination, usually made when in isolation.

FM 31-23 (ID)

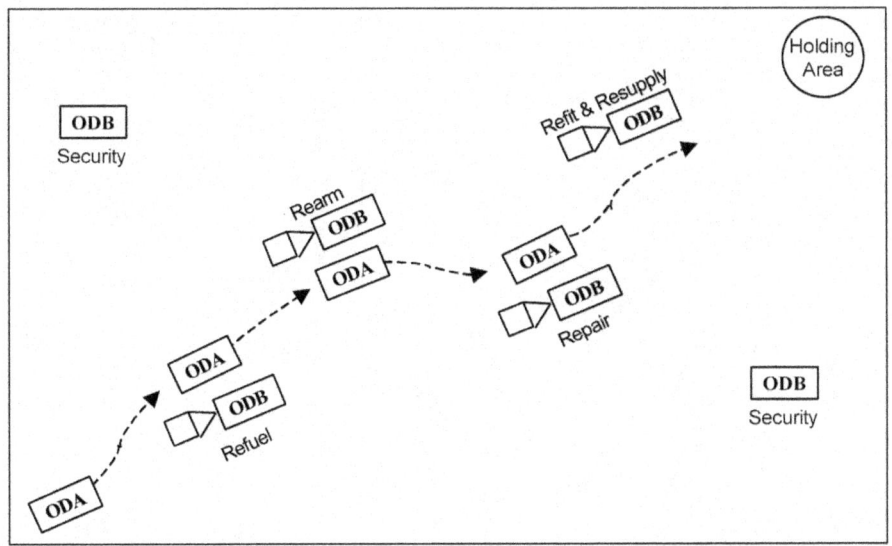

Figure 11-2. Example of a gas station or fluid MSS.

Resupply. The SFODA is resupplied with food and water. This is accomplished with the transfer of MRE boxes and 5-gal water jugs. The SFODA requests clothing and equipment replacement before the MSS is established.

Repair. The SFODB has limited ability to repair non-deadlined items of the SFODA's GMVs. Repair parts must be requested before the MSS is established.

MULTIPLE MSS CONCEPT

By using the SFODB to resupply SFODAs either by emplacing caches or by establishing an MSS, it is feasible to extend the operational range of one or more SFODAs. An SFODA, depending upon the mission, can infiltrate and move well past its operational range if it can be resupplied with fuel, at a minimum, to enable it to recover to friendly lines. The SFODA does not rely solely on an MSS for essential supplies. Plan for both an MSS and caches should be made.

The AOB-supported MSS is not a commonly used method of resupply. It is, however, an alternate method of resupply when Military Airlift Command (MAC) aircraft cannot support conventional methods (such as paradrops).

Depicted below are two SFODAs crossing a forward line of own troops (FLOT) or UWOA boundary and moving 750 mi to its target area. An AOB moves and establishes caches and an MSS to refuel the SFODAs. The AOB can be strengthened with the attachment of 5-ton trucks, mechanics, and supplies that the mounted detachment would not normally have access to on a mission (see Figure 11-3, page 11-4).

FM 31-23 (ID)

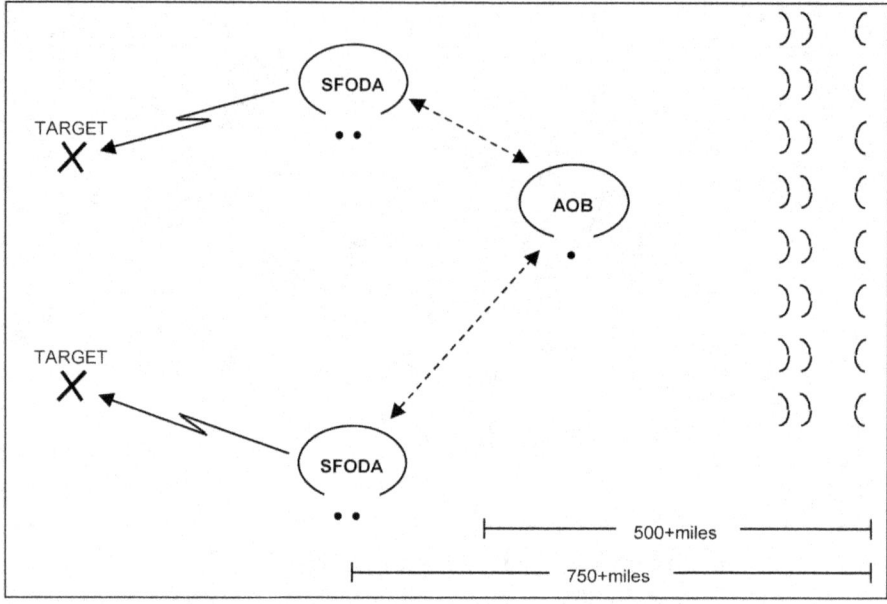

Figure 11-3. Forward-deployed AOB.

CACHING

During extended operations, a convenient means of resupplying mounted detachments is using caches. Mounted SFODBs or SFODAs could emplace caches as a secondary mission within their AO, to support mounted or dismounted missions of other SFODAs.

If possible, establish caches in an area before the enemy occupies it. Factors influencing the size and dispersion of caches are—

- Distance from the AOB, FOB, or launch site.
- Terrain.
- Enemy activity.
- Locally available resources.
- Type of operation(s) to be conducted.
- Distances and areas to be covered.
- Movement in cache areas.

The SFODA itself may be used to resupply dismounted elements. Mounted SFODAs can emplace many small caches that support dismounted elements or assisted evasion nets.

All caches have a main threat—detection. Select cache locations with concealment in mind. The contents of each cache should be as mixed as the operational requirements permit so that the destruction of any one cache will not create a shortage of any one commodity. A mixed content cache offers greater flexibility for use by other elements.

Cached fuel supplies must be emplaced so as not to contaminate the fuel. Both gasoline and diesel fuel, when stored for a long time, will auto-oxidize causing a breakdown of fuel components and render it unfit to use in military vehicles. Any fuel cached for more than six months will probably be unusable.

Before emplacing a long-term fuel cache, the detachment must treat the fuel to prevent auto-oxidizing and the growth of fungi. Military fuel is already treated to prevent auto-oxidizing but fuel procured overseas should be treated before caching. The detachment can get the procedures for treating fuel from the POL personnel assigned to their motor pool.

FM 31-23 (ID)

Chapter 12
Military All-Terrain Vehicle

The military all terrain vehicle (ATV) provides the commander another capability for his mounted detachment (Figure 12-1). This vehicle provides the detachment a highly mobile capability with additional cargo capacity the motorcycles do not have. The ATV is not simply a replacement for the military motorcycle; it is a different platform with characteristics and capabilities of its own.

Figure 12-1. Honda Foreman 300.

12-1

FM 31-23 (ID)

GENERAL

The ATV detachment's primary mission is long-range mounted SR. The ATVs provide the capability to conduct SR, DA, and UAR missions over a 10-day or 1,000-mi range without resupply in austere environments over difficult desert and jungle terrain.

The ATV detachment uses three primary vehicles: the GMV, the 6x6 ATV, and the 4x4 ATV. Although unarmed, the ATV gives the commander an additional tool to use when vehicles are necessary, but the large signature of four GMVs and two DOMs is undesired.

The military 4x4 ATV, with its internal fuel tank and two 5-gal fuel cans, can carry enough supplies to range 500 kilometers (km) or 5 days unassisted. With the larger 6x6 ATV, the detachment extends its range to 950 km, and with a HMMWV to carry additional supplies, the detachment's range extends to 1,500 km or 10 days without additional resupply.

ATV riders require extensive training to operate the vehicle safely and though all types of weather and terrain. Although small and powerful, its high center of weight makes it easy to flip if used incorrectly. Vehicle operators must know how to drive the vehicle properly, how to best load cargo to properly balance the vehicle, and have the training and capability to make on the spot repairs when deployed in the field far behind enemy lines.

The 6x6 ATV can carry twelve additional 5-gal fuel or water cans in its rear cargo box. The front cargo rack has a 75-lb capacity and 800-lb capacity for the rear cargo box.

The 4x4 ATV can carry four additional 5-gal fuel or water cans on its rear cargo rack. The front cargo rack has a 90-lb capacity and 180 lb for the rear cargo rack.

ORGANIZATION

The force package for an ATV mission is determined during mission planning. Unlike a motorcycle section, the ATVs are used full time. Like the motorcycle rider, an ATV operator will never operate alone.

Force Package 1—four 6x6 ATVs and eight 4x4 ATVs.

Package 1 offers the maximum flexibility for the commander. This configuration allows the element to operate as one element, or two, three, four, or six sub-elements. The 6x6 ATVs provide a mobile MSS to support the mission profile, while the 4x4 ATVs can range out ahead and around the detachment to accomplish the mission.

Each 6x6 ATV provides the needed cargo capacity for long-range "over-the-horizon" infiltration through all types of terrain. The eight 4x4 ATVs can also carry limited supplies, enabling them to operate up to 500 km with onboard supplies.

Mission essential equipment can easily be cross-loaded between the twelve ATVs. Rucksacks mount either to the front or rear cargo racks depending on mission profile and the detachment SOP. It is essential to secure any equipment by using small cargo straps or heavy duty flexible rubber straps. The front and rear cargo racks are open metal tubing and allow multiple points to hook straps. The rear cargo box on the 6x6 ATV also allows easy attachment of cargo straps and hooks.

FM 31-23 (ID)

Force Package 2—two GMVs with trailers and six 4x4 ATVs.

Package 2 offers the maximum in defensive firepower and cargo capacity for the ATV detachment. This configuration allows the detachment to operate as one element or two sub-elements. The two trailers provide the necessary additional cargo capacity for long-range "over-the-horizon" infiltration. Supplies can be mounted on the inside of the rear tire wells, against the rear side racks, and inside the trailers.

As with Force Package 1, each of the six 4x4 ATVs provide limited cargo capacity for infiltration through all types of terrain.

Force Package 3—one GMV with trailer and eight ATVs.

Package 3 combines the best aspects of Force Packages 1 and 2. The configuration gives the commander the capability to carry additional cargo, defensive firepower, and a vehicle package that can be easily transported by air or ground assets into the AO.

EMPLOYMENT CONCEPT

Operational employment of the ATV detachment is much the same as discussed earlier for the mounted detachment. Planners must consider the energy expended by the ATV riders when operating for long periods, especially at night. The planners must also consider the ATV detachment's reduced defensive firepower and its ability to move into and around an area with reduced vehicle and noise signatures. Planners must also consider that the GMVs use diesel fuel while the ATV uses MOGAS.

AIR INFILTRATION

The ATV can be sling loaded easily and rapidly by CH/MH-60s, CH/MH-53s, and CH/MH-47s. Additionally because of their relatively small size, they can easily be internally loaded on different types of aircraft (CH/MH-47, CH/MH-53, and the C/MC-130).

An MH-47 with an internal fuel bladder can carry six ATVs internally, without a fuel bladder it can fit up to eight. The MH-53 can carry six ATVs internally, and the MC-130 can carry one GMV and six ATVs.

MOVEMENT

Any type of movement begins with pre-movement planning of the routes, rally points, and navigational checkpoints. The ATVs uses only the first two levels of navigation, vehicle orienteering and DR, however, the ATVs can carry with them a GPS to check their navigation and confirm position. The rider carries on his person all maps and logs in case he has to separate from his vehicle rapidly.

Riding at night, particularly with NVGs, becomes very fatiguing. Operational burnout occurs between 3 to 6 hours with NVG use. The commander and ATV riders should be aware and plan for this.

REACTION DRILLS

The ATV is very vulnerable to small-arms fire and must, therefore, use its mobility and speed to distance itself from the enemy if contact is made.

If contact is made with the enemy, the ATV section tries to break contact by placing distance and cover between themselves and the enemy. The riders must be aware of each other, since if one ATV goes down, the others must gain position to support the downed rider until he can make his way to either an operational ATV or a covered or concealed position.

The ATV riders should be very adept at making quick stops, so they can effectively dismount their vehicle and gain the prone position if under a heavy volume of enemy fire.

EQUIPMENT

The ATV riders should carry mission-essential equipment and maintenance equipment. Listed below are the mandatory items of equipment for the ATV section, per rider:

- Individual weapon and LBE, to include ammunition, compass, first aid kit, water, strobe, flashlight, maps, and Kevlar helmet.

- SABER II radio with bone microphone, or other small RT (for emergency contact with main element).

- Pens flares and/or star clusters.

- ATV maintenance kit.

- Bug-out bag that includes food, survival, and comfort items.

FM 31-23 (ID)

Appendix A
M1025A2 (GMV) and M1114 (ARMORED) HMMWV

Mounted detachments will usually find themselves using the M1025A2 (GMV) but the M1114 provides an alternate form of transportation that has equally unique capabilities. The mounted detachment may determine one or the other vehicle better meets their specific mission requirements. This appendix explains the capabilities and provides a statistical comparison of both vehicles.

CAPABILITIES

M1025A2 (GMV) (Figure A-1). This vehicle is based on the army standard M1025A2, a scout vehicle. This vehicle is to scout platoons and military police units to replace aging M1026s now in use. The M1025A2 was modified to the GMV to focus on the accomplishment of one mission—long-range special reconnaissance in a desert environment. The GMV can conduct many other missions such as direct action, coalition support, humanitarian assistance, and peace enforcement.

Figure A-1. M1025A2 (GMV).

M1114 (armored) (Figure A-2, page A-2). This vehicle is based on the M1109. It is, however, extensively modified to protect the crew against 7.62-mm armor piercing (AP) ammunition and to protect against mine blasts (12 lb front, 4 lb rear). This vehicle is well suited for missions such as coalition support, humanitarian assistance, and peace enforcement. It is poorly suited for long-range special reconnaissance in a desert environment due primarily to its weight and carrying capacity.

FM 31-23 (ID)

Figure A-2. M1114 (armored) HWMMV.

OPERATIONAL CAPABILITIES

Weight and Payload. The M1114 (armored) has a very limited cargo carrying capability when compared to the M1025A2 (GMV). The curb weight of the M1114 is increased by 2,930 lb (a 1/3 increase) over the M1025A2. This added weight reduces the capability of the M1114 to negotiate loose or wet terrain. The payload of the M1114 was reduced by 930 lb (a decrease of 1/3) as compared to the M1025A2 (GMV). Another payload factor is the reduction of cargo space in the M1114. With a 1/3 decrease in load carrying capability and reduction is cargo space, the M1114 (armored) is restricted to operations of shorter duration than the M1025A2 (GMV).

Range and Duration. The M1025A2 (GMV) has a planning range of 500 mi or 10 days without a trailer and 1,000 mi or 15 days with a trailer. The M1114 (armored) has a planning range of 250 mi or 3 to 5 days.

Armor Protection. The M1114 (armored) offers 360-degree 7.62-mm AP protection. The M1025A2 (GMV) offers no such protection. The armor of the M1114 does not extend all the way to the rear to enclose the cargo area. An armor wall separates the cockpit from the cargo area.

STATISICAL COMPARISON OF M1025A2 (GMV) AND M1114 (ARMORED)

VARIANTS:	M1025A2 (GMV)	M1114 (ARMORED)
Curb weight	6,870 lb	9,800 lb
Payload	3,230 lb	2,300 lb
Gross weight	10,300 lb	12,100 lb
Engine (diesel)	6.5 liters	6.5 liters
Horsepower	160	190
Acceleration: 0-30 mph	8.2 sec	8.2 sec
0-50 mph	25.1 sec	25.1 sec
Maximum towed load	3,400 lb	4,200 lb
Cruising range (min)	320 mi	273 mi

Appendix B
Mission-Essential Task List

GMV MAINTENANCE

1-1. Conduct PMCS on GMV.
1-2. Implement the Lube Order.
1-3. Maintain OVM/basic issue items (BII).
1-4. Replace a half shaft.
1-5. Repair a flat tire.
1-6. Change a flat tire.
1-7. Replace the generator.
1-8. Change the long V-belt set.
1-9. Change the short V-belt set.
1-10. Replace hydraulic fan hose.
1-11. Replace the tie-rod end(s).
1-12. Replace lower radiator hose.
1-13. Replace the water pump.
1-14. Replace the glow plugs.
1-15. Repair seal on steering gearbox.
1-16. Repair loose or worn brake pads.
1-17. Troubleshoot the engine.
1-18. Troubleshoot the transmission.
1-19. Troubleshoot the drivetrain.
1-20. Troubleshoot the braking system.
1-21. Troubleshoot the steering system.
1-22. Troubleshoot the fuel system.
1-23. Maintain the WARN winch.
1-24. Conduct post-mission maintenance.
1-25. Troubleshoot the turret ring.

PLANNING AND INFILTRATION

2-1. Plan fuel usage for the mission.
2-2. Plan food and water usage for the mission.
2-3. Determine route distance by map.
2-4. Load the GMV.
2-5. Load the DOT.
2-6. Prepare vehicles for air movement.
2-7. Load vehicles in fixed-wing aircraft.
2-8. Load vehicles inside the helicopter.
2-9. Prepare vehicles for sling load.
2-10. Conduct sling load operations.
2-11. Prepare for sea movement.
2-12. Conduct sea movement.

MOUNTED OPERATIONS

3-1. Move in traveling formation.
3-2. Move in traveling overwatch.
3-3. Move in bounding overwatch.

3-4. Move in diamond formation.
3-5. Move in wedge formation.
3-6. Operate GMV in daylight.
3-7. Operate GMV using NVGs.
3-8. Operate GMV in MOPP 4.
3-9. Conduct a short duration halt.
3-10. Set up a laager.
3-11. Camouflage vehicles.
3-12. Conduct operations in a MSS.
3-13. Conduct a ferry mission.
3-14. Emplace and/or recover a cache.
3-15. Recover an immobile GMV without assistance.
3-16. Recover immobile vehicle with a GMV.
3-17. React to enemy ambush.
3-18. React to enemy chance contact.
3-19. React to enemy air attack.
3-20. React to indirect fire.
3-21. React to sniper fire.
3-22. React to laager site compromise.
3-23. Coordinate for passage of lines.
3-24. Interdict a target.
3-25. Evacuate wounded from a disabled GMV.

MOUNTED NAVIGATION

4-1. Navigate to 1/4 mi accuracy.
4-2. Complete a driver's log.
4-3. Employ dead reckoning.
4-4. Employ vehicle orienteering.
4-5. Conduct a square search.
4-6. Employ satellite navigation equipment.

MOUNTED WEAPONS

5-1. Employ the M-2 .50 cal HB MG.
5-2. Maintain the M-2 .50 cal HB MG.
5-3. Employ the MK-19.
5-4. Maintain the MK-19.
5-5. Employ the Stinger missile.
5-6. Maintain the Stinger missile.
5-7. Employ the UAS-11.
5-8. Maintain the UAS-11.

MOTORCYCLE MAINTENANCE

6-1. Conduct PMCS on the DOM.
6-2. Lubricate vehicle.
6-3. Maintain BII/safety equipment.
6-4. Replace a front tire,
6-5. Replace a rear tire,
6-6. Repair a flat tire,
6-7. Repair and/or replace wheel spokes.

6-8. Replace a drive sprocket.
6-9. Maintain a drive chain.
6-10 Replace a drive chain.
6-11. Replace capacitor discharge ignition.
6-12. Replace a spark plug and wire.
6-13. Replace a throttle cable assembly.
6-14. Replace a clutch cable assembly.
6-15. Adjust a clutch cable.
6-16. Replace a brake cable assembly.
6-17. Adjust a brake cable.
6-18. Maintain the coolant system.
6-19. Troubleshoot the coolant system.

MOTORCYCLE EMPLOYMENT

7-1. Operate the DOM during daylight.
7-2. Operate the DOM using NVGs.
7-3. Operate the DOM in MOPP 4.
7-4. Navigate to 1/4 mi accuracy.
7-5. Conduct a route reconnaissance.
7-6. React to an enemy ambush.
7-7. React to a chance contact.
7-8. React to an air attack.
7-9. React to indirect fire.
7-10. React to sniper fire.
7-11. Operate in a column formation.
7-12. Operate in a staggered formation.

Appendix C
Mounted Detachment Training and Evaluation Outline

This appendix contains the collective tasks, conditions, and standards for major collective tasks the mounted SFODA must perform to accomplish its primary wartime mission. These tasks, conditions, and standards are training and evaluation outlines (T&EOs). These T&EOs are tasks that support the accomplishment of the mounted SFODA's missions identified in this FM. This appendix gives the SFODA and higher commanders a tool to evaluate their mounted detachments. The T&EOs are training and evaluation tools to measure the performance of tasks that support critical wartime mission accomplishment. The task steps and performance measures that comprise these T&EOs are all considered critical to the accomplishment of the mounted detachment's wartime mission. A complete listing of the T&EOs is at Table 1.

The T&EO is used individually to train a single task or it is used in sequence with other T&EOs to train and evaluate larger tasks, an entire mission or series of missions, or internal and external evaluations.

This appendix is not found in a formal mission training plan (MTP) for an SFODA. The tasks listed here are primarily included as a guide to the commander in training his detachment and to supplement existing MTPs.

Table 1

TASK	TITLE	ELEMENT	PAGE
31-x-0100	Conduct Mission Planning (Mounted)	SFODA	1-2
31-x-0101	Conduct Pre-Mission Activities (Mounted)	SFODA	1-3
31-x-0102	Conduct Patrol Ferry Mission Planning (Mounted)	SFODA	1-4
31-x-0110	Infiltrate the Operational Area (Mounted)	SFODA	1-6
03-3-R313	Operate in an NBC Environment	SFODA	1-7
31-x-0200	React to Enemy Contact	SFODA	1-9
31-x-0201	Establish a Laager Site	SFODA	1-10
31-x-0202	Provide Logistics Support Within the AO	SFODB/SFODA	1-11

FM 31-23 (ID)

ELEMENT: SFODA

TASK: CONDUCT MISSION PLANNING (MOUNTED) (31-x-0100) (FM 101-5, FM 31-20, AND FM 31-23)

ITERATION	1	2	3	4	5	(circle)
COMMANDER/LEADER ASSESSMENT			T	P	U	(circle)

CONDITIONS: The SFODA has been alerted, received a mission briefing, and been placed in isolation to conduct mission planning for a mounted operation. This task should not be performed in MOPP4. This task is designed as a supplement to other existing mission planning tasks in formal ARTEP manuals.

TASK STANDARDS: The SFODA plans routes and load plans for a mounted mission to accomplish its assigned mission.

TASK STEPS AND PERFORMANCE MEASURES	GO	NO-GO

1. SFODA plans overland routes.
   a. Plans primary route.
   b. Selects tentative rally points on primary route.
   c. Selects tentative RAD sites on primary route.
   d. Plans emergency route(s).
   e. Selects tentative rally points on contingency/emergency route.
   f. Selects tentative RAD sites on contingency/emergency route.
* 2. Navigator prepares his navigation aids.
   a. Requests required maps.
   b. Prepares driver's logs.
   c. Briefs alternate navigators.
3. SFODA coordinates passage of lines (if applicable).
4. SFODA coordinates fire support plan (if available/applicable).
5. SFODA prepares generic vehicle load plan and packing list for each vehicle.
6. SFODA determines POL and PLL for duration of mission.
7. SFODA requests additional supplies, equipment, POL, and PLL.

* Indicates a leader task step.

TASK PERFORMANCE/EVALUATION SUMMARY BLOCK						
ITERATION	1	2	3	4	5	TOTAL
TOTAL TASK STEPS EVALUATED						
TOTAL TASK STEPS GO						
TRAINING STATUS GO/NO-GO						

ELEMENT: SFODA

TASK: CONDUCT PRE-MISSION ACTIVITIES (MOUNTED) (31-x-0101) (FM 101-5, FM 31-20, AND FM 31-23)

ITERATION	1	2	3	4	5	(circle)
COMMANDER/LEADER ASSESSMENT			T	P	U	(circle)

CONDITIONS: The SFODA has conducted initial planning and is in isolation to prepare for a long-range mounted mission. This task should not be performed in MOPP4. This task is designed as a supplement to other existing mission planning tasks in formal ARTEP manuals.

TASK STANDARDS: The SFODA loads and prepares its equipment IAW mounted mission.

TASK STEPS AND PERFORMANCE MEASURES	GO	NO-GO

1. SFODA receives requested supplies, equipment, POL, and PLL.
2. SFODA performs vehicle maintenance checks.
3. SFODA cleans and inspects weapons systems.
4. SFODA inspects and loads vehicle OVM/BII.
* 5. Navigator inspects and loads observation aids (binoculars, PVS-7s, and other).
6. SFODA installs communications equipment.
   a. Checks equipment for proper operation.
   b. Loads frequencies and frequency hopping crypto.
   c. Loads and tests secure crypto.
7. SFODA loads vehicles IAW mission load plan and detachment SOP.
8. SFODA cross-loads supplies and equipment between sections.

* Indicates a leader task step.

TASK PERFORMANCE/EVALUATION SUMMARY BLOCK						
ITERATION	1	2	3	4	5	TOTAL
TOTAL TASK STEPS EVALUATED						
TOTAL TASK STEPS GO						
TRAINING STATUS GO/NO-GO						

FM 31-23 (ID)

ELEMENT: SFODA

TASK: CONDUCT PATROL FERRY MISSION PLANNING (MOUNTED)(31-x-0102)
(FM 101-5, FM 31-20, AND FM 31-23)

ITERATION	1	2	3	4	5	(circle)
COMMANDER/LEADER ASSESSMENT			T	P	U	(circle)

CONDITIONS: The SFODA has been alerted, received a mission briefing, and been placed in isolation to conduct mission planning for a mission to ferry another SFODA into the AO. Maximum coordination will be established between the two detachments. This task should not be performed in MOPP4. This task is designed as a supplement to other existing mission planning tasks in formal ARTEP manuals.

TASK STANDARDS: The SFODA plans jointly with another SFODA for a patrol ferry mission to accomplish its assigned task.

TASK STEPS AND PERFORMANCE MEASURES	GO	NO-GO

1. SFODA establishes chain of command for patrol ferry mission with the dismounted element.
2. SFODA plans overland routes with dismounted element.
   a. Plans primary route.
   b. Selects tentative rally points on primary route.
   c. Selects tentative RAD sites on primary route.
   d. Plans contingency/emergency route(s).
   e. Selects tentative rally points on contingency/emergency route.
   f. Selects tentative RAD sites on contingency/emergency route.
* 3. Navigator prepares his navigation aids.
   a. Requests required maps.
   b. Completes driver's logs.
   c. Briefs alternate navigators.
4. SFODA coordinates passage of lines (if applicable).
5. SFODA coordinates fire support plan (if available/applicable).
6. SFODA plans drop-off points (primary and alternate).
7. SFODA develops and rehearses drop-off procedures with the dismounted element.
8. SFODA teaches and rehearses mounted reaction drills with the dismounted element.
9. SFODA plans cache locations and cached items to support dismounted element (if applicable).
10. SFODA develops and rehearses contact plans and procedures for the linkup (if applicable) with the dismounted element.
11. SFODA establishes a load plan for the dismounted element and cross-loads the dismounted element to the fullest extent possible.
12. SFODA briefs and rehearses mounted movement SOPs with the dismounted element.
13. SFODA maintains isolation for follow-on missions (OPSEC).

* Indicates a leader task step.

TASK PERFORMANCE/EVALUATION SUMMARY BLOCK						
ITERATION	1	2	3	4	5	TOTAL
TOTAL TASK STEPS EVALUATED						
TOTAL TASK STEPS GO						
TRAINING STATUS GO/NO-GO						

FM 31-23 (ID)

ELEMENT: SFODA

TASK: INFILTRATE THE OPERATIONAL AREA (MOUNTED) (31-x-0110) (FM 7-7, FM 17-15, FM 21-26, AND FM 31-23)

ITERATION	1	2	3	4	5	(circle)
COMMANDER/LEADER ASSESSMENT			T	P	U	(circle)

CONDITIONS: The detachment's operational area is located in hostile territory. Deploying detachment members, vehicles, and equipment have been transported to a launch site near the operational area within friendly territory. This task should not be performed in MOPP4. This task is designed as a supplement to other existing mission planning tasks in formal ARTEP manuals.

TASK STANDARDS: The SFODA infiltrates the objective area without compromising the mission.

TASK STEPS AND PERFORMANCE MEASURES	GO	NO-GO

* 1. SFODA commander directs movement from the launch base to SP.
    a. Ensures all changes to operation order (OPORD) are disseminated to detachment members.
    b. Confirms movement route provides concealment from enemy observation.
    c. Adjusts route(s) as necessary based on current METT-TC.
    d. Adjusts movement techniques/formations as necessary based on current METT-TC.
 2. SFODA makes SP time.
 3. SFODA conducts passage of lines (if available/applicable).
 4. SFODA maintains security.
    a. Employs active and passive counter tracking measures.
    b. Ensures rate of movement does not violate security.
    c. Takes actions at danger areas IAW detachment SOP.
    d. Maintains movement discipline (light, noise, litter, and interval).
    e. Mans weapons systems during movement (minimum of two vehicles' weapons systems manned at all times).
* 5. Navigator determines position within 1/4 mile at all times.
 6. SFODA keeps all personnel informed of its position during halts.

* Indicates a leader task step.

TASK PERFORMANCE/EVALUATION SUMMARY BLOCK						
ITERATION	1	2	3	4	5	TOTAL
TOTAL TASK STEPS EVALUATED						
TOTAL TASK STEPS GO						
TRAINING STATUS GO/NO-GO						

FM 31-23 (ID)

ELEMENT: SFODA

TASK: OPERATE IN AN NBC ENVIRONMENT (03-3-R313) (FM 3-4, FM 3-3, and FM 3-100)

ITERATION	1	2	3	4	5	M	(circle)
COMMANDER/LEADER ASSESSMENT			T	P	U	(circle)

CONDITIONS: The SFODA has a mission that requires it to cross and operate or continue to operate in an NBC contaminated area. The SFODA has individual and unit-organic NBC defensive equipment. Some iterations should be performed in MOPP4. This task is designed as a supplement to other existing mission planning tasks in formal ARTEP manuals.

TASK STANDARDS: The SFODA conducts its mission without sustaining NBC casualties.

TASK STEPS AND PERFORMANCE MEASURES	GO	NO-GO

* 1. The SFODA commander (with assistance from the operations and senior medical sergeants) considers additional environmental factors for operations in a contaminated area consistent with the assigned mission and the METT-TC.
    a. Uses all available NBC reports and intelligence to assess the contamination hazard.
    b. Determines the duration of the chemical agent hazard and/or establishes the operational exposure guidance (OEG).
    c. Selects routes to minimize exposure to the hazard consistent with the agent type, the duration of the hazard, or the OEG.
    d. Develops a plan for monitoring the hazard, using modified chemical or radiological survey techniques (point, flank, rear chemical sentry, or radiological monitor).
 2. The SFODA prepares to operate in a contaminated area.
    a. Assumes the correct MOPP level for the chemical or biological hazard.
    b. Protects exposed skin and uses dust masks (field-expedient or protective mask) for a radiological hazard.
    c. Employs the monitoring plan, detecting the cont amination hazard as it is encountered.
    d. Prepares vehicles and equipment for operations in a contaminated area.
 3. The SFODA conducts operations in a contaminated area consistent with the METT-TC.
    a. Continues to monitor activities.
    b. Avoids stirring up dust.
    c. Avoids dust clouds by increasing intervals between personnel, elements, or units for a radiological hazard.
    d. Avoids low ground and other contamination contact danger areas (chemical, biological, or radiological hazard).
    e. Identifies "clean" areas that may be used for MOPP gear exchange or temporary relief from MOPP4.

f. Remains aware of the quickest route and distance out of the contaminated area for emergencies.
   g. Conducts the mission and exits the area as quickly as possible without becoming an NBC casualty or violating the OEG.
4. The SFODA exits the contaminated area.
   a. Checks for chemical or radiological contamination.
   b. Identifies, treats, decontaminates, and prepares casualties for movement.
   c. Decontaminates to reduce the sp read of contamination (as required).
   d. Computes, records, and reports total dose.
   e. Submits NBC reports IAW the OPORD.
   f. Continues the mission IAW the OPORD.

* Indicates a leader task step.

TASK PERFORMANCE/EVALUATION SUMMARY BLOCK							
ITERATION	1	2	3	4	5	M	TOTAL
TOTAL TASK STEPS EVALUATED							
TOTAL TASK STEPS GO							
TRAINING STATUS GO/NO-GO							

FM 31-23 (ID)

ELEMENT: SFODA

TASK: REACT TO ENEMY CONTACT (31-x-0200) (FM 7-7, FM 17-15, FM 21-26, FM 31-23 AND UNIT SOP)

ITERATION	1	2	3	4	5	(circle)
COMMANDER/LEADER ASSESSMENT			T	P	U	(circle)

CONDITIONS: The SFODA is conducting a mounted operation and makes contact with enemy elements. This task should not be performed in MOPP4. This task is designed as a supplement to other existing mission planning tasks in formal ARTEP manuals.

TASK STANDARDS: The SFODA recognizes enemy contact or deliberate attack and employs its standard reaction drill to break enemy contact.

TASK STEPS AND PERFORMANCE MEASURES	GO	NO-GO

1. SFODA conducts immediate action drill to break enemy contact during one of the following scenarios IAW with detachment SOP and/or mission OPORD.
   a. Chance contact.
   b. Far ambush.
   c. Near ambush.
   d. Indirect fire attack.
   e. Attack from aircraft.
   f. Sniper attack.
   g. Laager site compromise.
2. SFODA conducts reconsolidation at contact location or designated rally point.
3. SFODA employs contingency route and procedures, if necessary.

TASK PERFORMANCE/EVALUATION SUMMARY BLOCK						
ITERATION	1	2	3	4	5	TOTAL
TOTAL TASK STEPS EVALUATED						
TOTAL TASK STEPS GO						
TRAINING STATUS GO/NO-GO						

FM 31-23 (ID)

ELEMENT: SFODA

TASK: ESTABLISH A LAAGER SITE (31-x-0205) (FM 7-7, FM 17-15, FM 21-26, FM 31-23 AND UNIT SOP)

ITERATION	1	2	3	4	5	(circle)
COMMANDER/LEADER ASSESSMENT			T	P	U	(circle)

CONDITIONS: The SFODA establishes a laager site for one period of daylight. This task should not be performed in MOPP4. This task is designed as a supplement to other existing mission planning tasks in formal ARTEP manuals.

TASK STANDARDS: The SFODA establishes a laager site that provides effective camouflage, observation, vehicle maintenance, and planning.

TASK STEPS AND PERFORMANCE MEASURES	GO	NO-GO

1. SFODA halts, emplaces security, and conducts a reconnaissance for the preplanned or a suitable laager site.
* 2. Operations sergeant (or another detachment member IAW the team SOP) sites the vehicles after the laager site is selected.
3. SFODA conducts a listening period once all vehicles are emplaced.
4. One section camouflages its vehicle while the other pr ovides security by manning the onboard weapons systems.
* 5. Navigator determines exact position that is disseminated to all SFODA members.
6. SFODA sterilizes entry tracks into laager site.
7. SFODA refuels and performs maintenance checks on its veh icles.
8. SFODA emplaces laager site defense systems (fighting positions, mines, early warning devices), if necessary.
9. Members perform weapons system maintenance.
10. Members take care of personal hygiene and observe crew rest with appropriate security.

* Indicates a leader task step.

TASK PERFORMANCE/EVALUATION SUMMARY BLOCK						
ITERATION	1	2	3	4	5	TOTAL
TOTAL TASK STEPS EVALUATED						
TOTAL TASK STEPS GO						
TRAINING STATUS GO/NO-GO						

FM 31-23 (ID)

ELEMENT: SFODB/SFODA

TASK: PROVIDE LOGISTICS SUPPORT WITHIN THE AO (31-x-0202) (FM 7-7, FM 17-15, FM 21-26, FM 31-23 AND UNIT SOP)

ITERATION	1	2	3	4	5	(circle)
COMMANDER/LEADER ASSESSMENT			T	P	U	(circle)

CONDITIONS: The SFODB/SFODA establishes an MSS, either laager or fluid type, to support another mounted detachment. This task should not be performed in MOPP4. This task is designed as a supplement to other existing mission planning tasks in formal ARTEP manuals.

TASK STANDARDS: The SFODB/SFODA safely conducts resupply of another mounted detachment.

TASK STEPS AND PERFORMANCE MEASURES	GO	NO-GO

1. SFODB/SFODA establishes an MSS position (laager or fluid type) IAW laager site employment procedures and standards.
2. SFODB/SFODA establishes contact or linkup with element to be serviced or resupplied.
3. The SFODB/SFODA resupplies, refuels, rearms, refits, and provides maintenance support to the element. Exact services depend on need.
4. If the serviced element remains through a period of daylight, the SFODB/SFODA develops an extended laager site and the element integrates into the perimeter.
5. Upon completion of the MSS, the SFODB/SFODA vacates the site and sterilizes the area.

TASK PERFORMANCE/EVALUATION SUMMARY BLOCK						
ITERATION	1	2	3	4	5	TOTAL
TOTAL TASK STEPS EVALUATED						
TOTAL TASK STEPS GO						
TRAINING STATUS GO/NO-GO						

Appendix D
Mounted Detachment Training Program

PURPOSE: To train SFODA(s) in the operation and use of the ground mobility vehicle (GMV) and/or HMMWVs for DA and/or SR missions.

SCOPE: SFODA plans for and conducts cross-country movement to accomplish the DA and/or SR mission.

SPECIAL INFORMATION: Instruction will cover two and a half days and nights. Eight hours in class at a field site, a five-hour planning and packing practical exercise (PE), and an eighteen-hour cross-country PE.

SECURITY CLEARANCE: For Official Use Only.

DATA:

Course Length	54 hours
Maximum Class Size	6 SFODAs
Minimum Class Size	1 SFODA

TRAINING START DATE: TBD

TRAINING DEVELOPMENT PROPONENT: 5th Special Forces Group (Airborne), Fort Campbell, Kentucky 42223-5000

REMARKS: Training is based on the SFODA(s) to be trained having vehicle (HMMWV) assets.

COURSE SUMMARY

COURSE: Special Forces Mounted Operations

HOURS:

Academic Time	14.0
Meal/Commander's Time	12.0
Practical Exercise	28.0

TOTAL COURSE HOURS: 54.0

HOURS BY SECURITY CLASSIFICATION:

FOUO   54.0

CLASS SIZE:

Maximum	6 SFODAs
Minimum	1 SFODA

FM 31-23 (ID)

POI FILE INDEX

COURSE: Special Forces Mounted Operations

POI FILE NUMBER	TITLE	HOURS	ANNEX
57600	PMCS Modified M1025A2 HMMWV (GMV)	4	A
57601	Introduction to SF Mounted Operations	1	A
57602	Mission Planning/Load Plans	3	A
57603	Land Navigation/Duties of the Navigator	1	A
57604	Formations and Movement	7	A
57605	Laager Sites/Camouflage	3	A
57606	Mission Planning PE	5	A
57607	Cross-Country Movement PE	18	A

ANNEX A (SUMMARY OF INSTRUCTION) - POI SF Mounted Operations

COURSE: Special Forces Mounted Operations

POI FILE NUMBER	TITLE	SECURITY CLASSIFICATION	HOURS BY TYPE
57600	PMCS Modified M1025A2 HMMWV (GMV)	FOUO	1 C; 3 PE
	Classes cover correct before, during, after, and post-operations checks on the modified M1025A2 HMMVW (GMV).		
57601	Introduction to SF Mounted Operations	FOUO	1 C
	Classes will consist of a basic introduction to the capabilities of GMV teams and the advantages and disadvantages of the GMVs.		
57602	Mission Planning/Load Plans	FOUO	3 C
	Training will cover mission planning for a long-range cross-country movement, including fuel, PLL, food, and water needed.		
57603	Land Navigation/Duties of the Navigator	FOUO	1 C
	Training will cover the differences between mounted and dismounted navigation and the duties of the detachment/vehicle navigators.		
57604	Formations and Movement	FOUO	7 PE
	Training will cover basic movement techniques and formations to include PE (with NVGs).		
57605	Laager Sites/Camouflage	FOUO	3 C
	Training will demonstrate laager sites and methods to camouflage equipment.		
57606	Mission Planning PE	FOUO	5 PE
	Students will be given a route to plan for and will pack their HMMWVs for an overnight PE.		
57607	Cross-Country Movement PE	FOUO	18 PE
	Students will move cross-country to checkpoints and finally to a RAD site where they will laager and camouflage their HMMWVs.		

ANNEX B (EQUIPMENT/RESOURCE SUMMARY) - POI SF Mounted Operations

COURSE: Special Forces Mounted Operations.

Requirements for Training

===

Each detachment attending training should have at least one vehicle per 4 men (1:3 preferred).

Each vehicle will need 5 diesel fuel and 5 water cans.

Each vehicle will need at least one desert camouflage net with pole set.

We will require a parade-type field with bleachers to conduct training and to use as a staging base during the overnighter.

ANNEX C (PROJECTED TRAINING OUTLINE) - POI SF Mounted Operations

COURSE: Special Forces Mounted Operations

Day 1

0700-0800   PMCS Modified M1025A2 HMMWV (GMV)

0800-0900   Introduction to SF Mounted Operations

> Classes will consist of a basic introduction to the capabilities of GMV teams and the advantages and disadvantages of the GMVs.

0900-1200   Mission Planning/Load Plans

> Training will cover mission planning for a long-range cross-country movement including fuel, PLL, food, and water needed.

1200-1300   Chow

1300-1400   Land Navigation/Duties of the Navigator

> Training will cover the differences between mounted and dismounted navigation and the duties of the detachment and vehicle navigators.

1400-1800   Formations and Movement

> Training will cover basic movement techniques and formations, including some PEs.

1800-1900   Chow

1900-2200   Formations and Movement

> Training will cover basic movement techniques and formations, including some PEs (with NVGs).

2200-0000   Commander's Time

Day 2

0000-0700   Commander's Time

0700-0800   PMCS Modified M1025A2 HMMWV (GMV)

0800-1100   Laager Sites/Camouflage

> Training will show how to select and camouflage laager sites.

1100-1600	Mission Planning PE

Students will be given a route to plan for and will pack their HMMWVs for an overnight PE.

1600-0000	Cross-country movement PE

Students will move cross-country to checkpoints and finally to a RAD site where they will laager and camouflage their HMMWVs.

Day 3

0000-0600	Cross-country movement PE (continued)
0600-1000	Return to SP/after-action review

Students will return to the SP, perform PMCS on their vehicles, and conduct an after-action review.

1000-1200	Post-operations PMCS on Modified M1025A2 HMMWV (GMV)

FM 31-23 (ID)

Appendix E
Pre-Mission Checklist

Use the following pre-mission checklist to conduct a vehicle readiness inspection before operations, either at the launch site, FOB, or staging area.

Detachment Commander—

- Keeps the detachment informed of times, locations, and changes.
- Ensures personnel perform their respective duties.
- Checks that assigned and attached personnel understand the mission completely.

Operations Sergeant—

- Verifies that the vehicles are loaded IAW the established load plans.
- Collects and inspects all pre-mission maintenance reports.
- Makes sure all special equipment is loaded.
- Checks that each detachment member is aware of contingency plans during movement and halts.
- Ensures all timelines are met.

Vehicle Navigator—

- Assures all required maps are present and loaded.
- Checks for functional operation of compasses, odometers, speedometers, and satellite navigation equipment.
- Verifies that radios are installed, loaded with correct frequencies and crypto keys, and work properly.
- Checks that all additional equipment (POL, PLL, BII, batteries, extra radios, and optical equipment) is loaded properly.
- Ensures that his route-planning log is prepared.

Vehicle Primary Operator—

- Ensures vehicle is mechanically ready for operation.
- Verifies all OVM, BII, manuals, and vehicle common equipment are loaded.
- Checks that the vehicle is topped off with fuel and that all fuel and water cans are full and loaded on the vehicle.

- Ensures all additional rations are loaded on the vehicle.

Vehicle Weapons System Operator—

- Checks that the weapon functions properly.
- Ensures ammunition load is onboard.
- Checks that the internal load is squared away and all equipment/supplies are loaded properly.
- Ensures the vehicle camouflage system is installed and that all components are serviceable and present.

Motorcycle/ATV Operator—

- Ensures motorcycle is functionally operational.
- Checks that motorcycle PLL and POL are onboard.

Assistant Detachment Commander—

- Verifies that all detachment POL/PLL is loaded.
- Ensures all special tool kits, general mechanics' tool kits, plug kits, and other non-common vehicle equipment are loaded.
- Checks that all vehicles are loaded IAW with detachment SOP and vehicle loading plan(s).
- Verifies that all assigned and attached personnel know the entire mission plan including infiltration, exfiltration, escape and evasion plan, routes, and all contingencies.

Appendix F
Load List

This appendix contains a recommended load list of POL, PLL, ammunition, and subsistence items required for the mounted detachment.

GENERIC MISSION PROFILE
(10 days or 1,000 mi using four GMVs, two DOMs, and two DOTs)

GMV-Related Parts and Equipment:

Spare tire complete	4 each
Spare tire, without rim	2 each
Long half-shaft	2 each
Short half-shaft	2 each
High lift jack with handle	2 each
Tie-rod end	2 each
Ball joint	2 each
Lower radiator hose	2 each
Oil filter	2 each
Fuel filter	2 each
Generator with key and pulley	1 each
Oil pan plug	2 each
Long V-belt set	2 each
Short V-belt set	2 each
Fuel cap	2 each
10 B&C rated fire extinguisher	4 each
200-psi air compressor	4 each
GM tool kit with metric supplement	2 each
Torque wrench	2 each
Tire repair kits	2 each
Tire plug kit	4 each
Slave cable	2 each
Spout, fuel flexible	6 each
Cargo general utility (CGU)-1/B tie-down strap	12 each
Motorcycle tie-down straps	8 each
Grease gun with flexible hose	2 each
Hydraulic fan hose with fitting	2 each
Tow bar	1 each
Tow straps	4 each

Motorcycle Parts and Related Equipment:

Air filter element	1 each
Oil filter element	2 each
Spark plug with spare wire	4 each
Front inner tube	3 each
Rear inner tube	3 each
Front tire complete with tube rim	1 each

Rear tire complete with tube rim	1 each
Brake handle assembly	2 each
Clutch handle assembly	2 each
Throttle assembly	1 each
Tachometer/speedometer cable	1 each
Clutch cable	2 each
Brake cable	2 each
Spare spokes, front	5 each
Spare spokes, rear	5 each
Foot pegs, set	1 each
Shifter lever	1 each
Brake pedal	1 each
Drive chain	2 each
Rear sprocket	1 each
Handlebar	1 each
Ignition element	1 each

POL Products:

Diesel fuel, DF-2 (not including onboard fuel tanks)	360 gallons (72 cans)
MOGAS (not including onboard MC tanks)	40 gallons (8 cans)
Oil, 15w-40, quart cans	24 cans/bottles
Oil, 90w, pint bottles	8 bottles
Oil, fork, pint bottles	2 bottles
Brake fluid, quart can	4 cans
Dextron II, transmission fluid	16 cans (quarts)
Grease automotive and artillery (GAA), in tubes for grease gun	2 tubes
Solvent, 1/2 gal can	2 cans
Window cleaner, 8 ounce bottle	4 bottles
Antifreeze, undiluted (gal bottle)	6 bottles
Chain lube, graphite, dry, tube	2 tubes
Fuel cans, 5 gallon	80 each

Subsistence:

Water, 5 gallon cans	200 gallons (40 cans)
MRE, case	20 cases

Ammunition, Demolitions, and Pyrotechnics:

A059 5.56-mm ball	5,040 rounds
A363 9-mm ball	540 rounds
A543 .50 cal 4+1 API-T	1,200 rounds
B542 40-mm linked MK 19	576 rounds
B534 40-mm multi projectile M203	12 rounds
B546 40-mm HE M203	24 rounds
H557 AT-4	16 each
K181 mine AT M21	8 each
K143 mine AP M18A1	12 each
K121 mine AP M14	40 each
M039 charge, demo shape 40 lb	4 each
G881 grenade, fragmentary M67	48 each

G900 grenade, incendiary ....8 each
G930 grenade, smoke HC ....16 each
PJ02 stinger SAM ....2 each

GENERIC MISSION PROFILE

(5 days or 500 mi using four GMVs, and two DOMs)

GMV-Related Parts and Equipment Same as above.

Motorcycle Parts and Related Equipment Same as above.

POL Products:

Diesel fuel, DF-2 (not including onboard fuel tanks) ....120 gal (24 cans)
MOGAS (not including onboard MC tanks) ....20 gal (4 CANS)
Oil, 15w-40, quart cans ....12 cans/bottles
Oil, 90w, pint bottles ....4 bottles
Oil, fork, pint bottles ....2 bottles
Brake fluid, quart can ....4 cans
Dextron II, transmission fluid ....8 cans (quarts)
GAA, in tubes for grease gun ....2 tubes
Solvent, 1/2 gal can ....2 cans
Window cleaner, 8 ounce bottle ....4 bottles
Antifreeze, undiluted (gal bottle) ....6 bottles
Chain lube, graphite, dry, tube ....2 tubes
Fuel cans, 5 gallon ....28 each

Subsistence:

Water, 5 gallon cans ....100 gallons (20 cans)
MREs, case ....10 cases

Ammunition, Demolitions, and Pyrotechnics Same as above.

NOTE: These load lists are meant as an aid in planning. Specific missions will dictate the essential load list for the detachment.

Appendix G
Fuel Estimation Formula

During mission preparation and planning, the team members use the following formula when estimating fuel requirements.

	_____	Total miles of mission (mission distance)	
divided by	_____	vehicle mpg average	light load, x country = 12 mpg heavy load, x country = 10 mpg light load, highway = 10 mpg heavy load, highway = 7 mpg fully loaded trailer = add 5 mpg
equals	_____	gal necessary per vehicle	
plus	_____	% of gal necessary added for map error	1:250,000 scale = 15% 1:100,000 scale = 10% 1: 50,000 scale = 5%
equals	_____	adjusted gal necessary per vehicle	
multiply by	_____	number of vehicles on mission	
equals	_____	gal necessary for detachment	
plus	_____	15% safety factor	
equals	_____	total detachment fuel requirements	
minus	_____	gal carried in vehicle fuel tanks (25 gal per vehicle tank)	
equals	_____	gal of fuel to be carried in 5 gal fuel cans	
divide by	_____	gal per can (U.S. fuel can = 5 gal)	
equals	_____	5-gal cans necessary to carry remaining fuel requirements	

FM 31-23 (ID)

Appendix H
Water Estimation Formula

During mission preparation and planning, the team members use the following formula when estimating water requirements.

	_____	number of detachment personnel
multiply by	_____	number of quarts per day (min 4-6 quarts)
multiply by	_____	number of days of mission duration
equals	_____	mission water requirements
plus	_____	15% safety factor
equals	_____	total water requirements
divide by	_____	gal per can (U.S. water can = 5 gal)

Appendix I
CH-47/MH-47 Internal Load Operations

This appendix addresses the procedures for preparing the GMV for internal load operations with a CH-47/MH-47 helicopter.

GENERAL

An MH-47 can lift one GMV at a time. Its lift capability is dependent on range, weather, and altitude. The vehicle and helicopter must be prepared before loading.

There is a two-inch clearance on each side and on the top of a GMV inside the aircraft.

The helicopter must land on a flat landing zone. If not flat, the MH-47's frame will bend and trap the GMV inside the aircraft.

The vehicle driver should rehearse loading and unloading using the same mission aircraft and aircrew member.

EQUIPMENT

The following equipment is required for internal load operations:

- 10,000-lb CGU straps or chains (provided by aircrew).
- Antenna tie-down.
- Small cargo straps for the motorcycles.

VEHICLE PREPARATION

The detachment personnel prepare the vehicles as follows—

- Remove the weapon system and weapon pintle.
- Close the gunner's hatch.
- Secure all loose cargo inside GMV.
- Remove the antenna or tie it down so that it is no higher than the latch of the turret hatch.
- Remove the mirrors or push them inward.
- Unlatch the trailer (if present).
- Ensure the navigator has cargo straps for the motorcycles (if present).

INTERNAL LOAD PROCEDURES

For an infiltration from a secure LZ—

- The MH-47 lands and the ramp is lowered and fully extended.

- The driver drives GMV 3' to 5' from the tail of the aircraft (A/C) facing away from rear ramp.

- An A/C crewmember exits through the ramp, moves to the left front (driver's side) of the GMV, and prepares to ground guide vehicle.

- The vehicle driver—

    - Aligns the GMV 3' to 5' behind MH-47 facing out.

    - Ensures mirrors are removed or pushed in.

    - Places the GMV in low lock drive.

    - Takes all instructions from A/C crewmember standing to the left front of the GMV.

    - Backs the GMV into the MH-47 and prepares to ride aircraft inside the GMV.

> **CAUTION**
> **Align the driver's door with the rear window to let him exit the aircraft in an emergency.**

- The navigator—

    - Ensures the antenna is tied down or removed.

    - Exits the GMV and positions himself to the right front of the vehicle.

    - Assists the A/C crewmember in loading the GMV.

        NOTE: Only the A/C crewmember ground guiding the GMV is allowed to give instructions to the driver. The navigator uses hand signals to pass information to the A/C crewmember who uses them to help him guide the diver into the helicopter.

    - Loads the A/C after the GMV is backed up the ramp.

    - Assists the A/C crew to secure 10,000-lb CGU strap.

    - Accounts for all vehicle personnel.

    - Gives the A/C crew chief thumbs up when loaded.

- The weapons operator—

- Ensures weapons and pintle are removed and secured in the GMV.
- Ensures the turret hatch is closed.
- Climbs into the A/C as soon as the ramp is lowered and before loading the GMV.
- Takes position with crew in the front of the A/C.
- Assists the A/C crew to secure the 10,000-lb CGU strap.

> **CAUTION**
> **DOM operators (if present) are entering the aircraft.**
> **Do not step in front of them while they are loading.**

- The DOM operators—
  - Load on the first A/C infiltrating. They provide initial security of the LZ.
  - Position themselves behind the A/C.
  - Wait while the crew and the GMV are loaded on the A/C.
  - Wait for the crew chief's or navigator's signal that the vehicle is loaded and secure.
  - Drive to the ramp and, for an infiltration, back the DOMs side by side on the ramp. Riders may need help from the navigator or crew chief to load.
  - Secure motorcycle to ramp using DOM tie-down straps.

For an emergency exfiltration—

- The GMV approaches the A/C from the rear and drives on forward.
- All duty positions remain the same.
- The DOMs approach the last A/C leaving and drive on forward.
- The team members may use chemlites to mark the sides of the ramp when loading onto the A/C under limited visibility using NVGs.

TIME WARNINGS

Actions at the time warnings are—

- Six Minutes—Wake all vehicle members.
- Four Minutes—Gunner unstraps front (of aircraft) cargo straps.

- One Minute—Driver starts engine (usually works better only 30 seconds out).

- Upon Landing—Navigator unstraps rear (of aircraft) cargo straps.

UNLOADING AIRCRAFT

The DOM operators drive off the A/C and provide security.

The navigator and the A/C crew chief exit ramp.

The crew chief ground guides the GMV off the A/C.

The gunner walks off the ramp.

The navigator and gunner climb aboard the GMV.

The gunner installs the weapon(s) as soon as possible.

FM 31-23 (ID)

Appendix J
Sling Load Operations

This appendix contains procedures for preparing the GMV for sling load operations when using Task Force 160 MH-47s.

Note: Prepare the GMV for movement IAW FMs 55-450-4 and 55-450-5.

GENERAL

The MH-47 has a multi-hook maximum lift capability of 25,000 lb that can be loaded on the fore and aft hooks as a tandem load.

The UH-60A has a hook tensile strength of 8,000 lb. However, its lift capability is dependent on factors such as temperature and altitude.

A combat-loaded GMV normally approaches or exceeds 9,000 lb, so it is recommended that the MH-47 be used for GMV sling load operations. The UH-60 can be used for trailer sling loads or unloaded GMVs.

Only a qualified rigger, pathfinder, or air assault trained individual may inspect the load for proper preparation and rigging.

The method of sling load operations as described in this appendix was developed jointly by TF-160 and the 5th SFG(A) during Operation Desert Shield. It was designed primarily to enable the pilots to quickly attach a sling load during limited to zero visibility and during "brownout" conditions that normally occur in the desert.

EQUIPMENT

The following equipment is needed for sling load operations:

- Cloth tape (100 mph tape).
- Cord (parachute line).
- Padding material (honeycomb), if available.
- 80-lb cord.
- ¾-inch plywood, 4' x 8' sheet (1 per vehicle).
- Sling set, helicopter, 25,000 lb capacity, NSN 1607-01-027-2900 (2 complete sets per GMV, plus an additional 25,000 lb clevis per sling set [8 legs and 4 clevises per GMV]).

GROUND MOBILITY VEHICLE PREPARATION

Prepare the GMV as follows—

- Secure all loose cargo.

- Take off or secure all doors (if left on the GMV, lower windows completely).

- Turn front tires so that they face straightforward.

- Remove and secure antennas and pad the antenna base.

- Remove weapons systems and pintle.

- Tape or pad all lights, including dashboard, to prevent flying glass.

- Secure hood latches with 2-inch green cloth tape.

- Remove or push in mirrors. If pushed in, secure them with 550 cord tied between each mirror and running through the inside front of the vehicle.

- Place an "X" of 2-inch green cloth tape on both sides of the windshield to make it shatterproof.

- Place transmission in neutral with brake off. (*Note: Different from Army Standard.*)

- Unlock and secure steering wheel with 550 cord to prevent wheels from turning.

- Tape key and lock together and secure inside the vehicle (if a lock is used).

- Secure the tailgate.

- Close all open faced hooks and shackles with tape or 550 cord to prevent them from working loose.

- Take off and secure pioneer tools inside the GMV, or secure them to rack with 550 cord.

- Secure the winch hook and close the open-faced hook with 550 cord.

- Pad GMV where the chains may rub against the vehicle.

SLING PREPARATION

Prepare the sling sets as follows—

- Use 2 sling sets for each GMV. Lay out both slings to one side of the vehicle (the side the helicopter is landing on) with a clevis (apex) and two legs running to an intermediate clevis with two more legs going to the attaching points on the GMV. When attached to the MH-47, the front sling set attaches to the helicopter's fore attaching hook and the rear set attaches to the helicopter's rear hook.

- Use the GMV's proper sling lifting points. The front two points are on top of the hood, and the rear two points are under the tailgate.

- Run the rear chains properly through the chain guides on the sides of rear tailgate.

- Use the proper grab link count. Grab link count on the GMV is 62 links in front and 34 links rear.

- Wrap the excess chain around the chain leg and secure it with 550 cord.

- Tighten and secure the apex securing pin nut and safety bolt on all four clevises with 2-inch green tape or cotter pins.

- Make two complete 360-degree inspections around the vehicle and from each lift point to apex.

NOTE: Only a qualified rigger, pathfinder, or air assault trained individual may inspect the load for proper preparation and rigging.

- Make sure that all personnel know and understand their responsibilities during helicopter operations.

HOOKUP PROCEDURES

CAUTION

Attach the sling set(s) to the correct attaching points on the helicopter. The front set to the front lifting point, and the rear set to the rear lifting point. This attachment allows the helicopter to carry the GMV safely and aerodynamically.

GMV moves to helicopter (preferred hookup method, especially under limited visibility conditions).

- After the aircraft lands, the driver moves the vehicle to 3 to 5 feet from the left side of the aircraft. Driver then secures the vehicle, ensures the GMV is in neutral, the brake is off, and the wheels are straightforward.

- The navigator secures the apex of the front sling set and attaches it to the front lift hook of the aircraft.

- The gunner secures the apex of the rear sling set and attaches it to the rear lift hook of the aircraft.

Vehicle stays in a stationary position (less preferred hookup method).

- The driver moves to the front and to the side of the GMV about 50 meters and prepares to guide the helicopter in (coordinate this action with the aircrew, TF-160 pilots normally do not need or want a ground guide when landing). The pilot lands the MH-47 3 to 5 feet from the GMV on the preplanned side (normally the right side of the vehicle).

- The navigator secures the apex of the front sling set and kneels in front of the GMV's front bumper on the side the helicopter is landing. After the helicopter lands, the navigator moves to the helicopter and attaches the sling set to the front lift hook of the aircraft.

- The gunner secures the apex of the rear sling set and kneels behind the GMV's rear bumper on the side the helicopter is landing. After the helicopter lands, the navigator moves to the helicopter and attaches the sling set to the rear lift hook of the aircraft.

LOADING PROCEDURES

Personnel follow the following loading procedures:

- Once GMV is hooked up, the crew of the vehicle moves to the rear of the MH-47 and enters the aircraft. If motorcycles are being used, they are loaded on the helicopter at this time.

> **CAUTION**
> **DOM operators (if present) are entering aircraft.**
> **Do not step in front of them while loading.**

- The riders wait for the crew chief's or navigator's signal to drive the DOMs up the helicopter ramp. Drivers may need help from the navigator or crew chief at this time. The DOMs are secured using DOM tie-down straps.

TAKEOFF AND LIFTING

Once the vehicles and personnel are loaded, the vehicle navigator signals the crew chief and the pilot lifts off, swings over the top of the GMV, and lifts the load. The aircraft then proceeds to its destination.

RELEASING THE LOAD

- At the release LZ, the pilot maneuvers the aircraft into position to lower the GMV while the loadmaster watches through the bottom of the helicopter.

- The loadmaster informs the pilot when to cut the load free. After being cut free, the GMV touches the ground and rolls forward for a short distance.

NOTE: *TF-160 pilots normally release the GMV before it fully touches the ground and while the helicopter is still moving forward.*

- After GMV is cut loose, the pilot lands the aircraft by swinging to the right of the vehicle.

- If present, motorcycles exit the aircraft and the riders secure the LZ.

- The vehicle crew exits the helicopter and waits at the rear for it to takeoff.

- The driver quickly inspects the vehicle, releases the steering wheel, and prepares to depart the LZ.

- The navigator secures the front sling set on vehicle and prepares for movement.

- The gunner secures the rear sling set on vehicle and prepares for movement.

- Detachment carries out its assembly plan and moves off the LZ. As soon as possible, the detachment will stop, mount the weapons systems, and repack the sling sets for movement.

Appendix K
Motorcycle Training Program

When used correctly, the motorcycle is the most valuable tool the detachment has in its inventory. However, safe operation of the motorcycle requires extensive training and rehearsal. Even soldiers with years of experience riding dirt bikes will find that operating a military motorcycle at night, with NVGs, is far different than what he is used to. This appendix presents a recommended motorcycle program of instruction (POI) that has been used extensively by the 5th SFG(A).

0001

MOTORCYCLE TRAINING PROGRAM
COURSE OUTLINE

1. LESSON PURPOSE. The purpose of this class is to enable the SF soldier to operate a DOM in rough terrain, through any weather, on any night or day tactical mission.

2. OBJECTIVES.

    a. The attached POI is provided to train detachment members on a military motorcycle, primarily in a desert environment. It is laid out in a logical progression that may be extended or condensed, based on the entry skill level of the selected riders. It is designed to instruct a minimum of two riders or a maximum of twelve riders.

    b. The hours programmed for each item of hands-on training have maintenance already programmed in. If possible, riders should attend a certified maintenance course. This course would greatly improve each rider's ability to identify and perform operator emergency maintenance.

    c. This POI is based on the following rider prerequisites:

        1. Prior experience, civilian or military, on some type of motorcycle.

        2. Current civilian motorcycle license. Rider must have a civilian motorcycle license to obtain a military motorcycle license.

        3. Current DA Form 348.

        4. Good physical condition with no profiles.

        5. Highly motivated, mature, and mechanically inclined individual.

MOTORCYCLE PROGRAM OF INSTRUCTION

4. PHASE I. CLASSROOM INSTRUCTION   14.5 TOTAL HOURS

   a. Motorcycle components, controls, and operator maintenance. (2 hr)

      Operator's manual overview using pertinent motorcycle service manual.

   b. Safety and proper wearing of equipment. (1/2 hr)

      Explaining the various items that must be worn while operating the motorcycle.

   c. Basic riding skills. (2 hr)

      Centering, riding posture, starting and stopping, weight transfer, traction, body position for balancing, use of feet and foot pegs, acceleration, shifting and braking.

   d. Advanced riding skills.

      1. Braking: bumps, turns, and slides. (1 hr)

      2. Turning: slides, sweeps, tight, cambered, down, up, banked, and S-turns. (2 hr)

      3. Jumping: minimize and maximize distance, ditch, downhill, uphill, low, fall away, and stepped jumps. (2 hr)

      4. Terrain techniques: climbing, descending, downhill braking, uphill throttle control, sand, mud, and water. (2 hr)

   e. Tactical employment of the DOM. (1 hr)

      Brief overview of the employment of the DOM on a tactical mission.

   f. Performing troubleshooting techniques on the motorcycle. (2 hr)

      Explaining different troubleshooting techniques that may occur during tactical mission operations.

4. PHASE II: BASIC SKILLS (demo and practical exercise)   4 TOTAL HOURS

   Basic riding techniques.

      Starting, stopping, acceleration, shifting, braking, and direction changes. (4 hr)

5. PHASE III: ADVANCED SKILLS (demo and practical exercise)   15+ TOTAL HOURS

   a. Braking skills. (2 hr)

      Bumps, turns, and slides.

b. Turning. (4 hr)

   Slides, sweeps, tight, cambered, down, up, banked, and S turns.

c. Jumping. (4 hr)

   Minimize and maximize distance, ditch, downhill, uphill, fall away, and stepped jumps.

d. Terrain techniques. (5 hr)

   Climbing, descending, downhill braking, uphill and downhill throttle control, sand, mud, and water.

e. Team IADs.

   Team SOP.

f. Nighttime riding. (at least 2 hr per student)

   Using all skills taugh t.

6. PHASE IV: EVALUATION.

   Execute phases I, II, and III under supervision to mission standards.

Appendix L
Post-Operations Maintenance Procedures

This appendix provides a recommended Post-Operations Maintenance and Recovery (POMR) SOP for use by units in garrison and the field. Use these procedures upon the completion of all missions to include field training exercises, emergency deployment readiness exercises, Army Training and Evaluation Programs, or other operational missions. It is too easy to just "make do." Adherence to this SOP will ensure equipment reliability and detachment readiness.

CONCEPT OF POMR

POMRs consist of actions and deadlines to return the detachment to a state of mission readiness, by accomplishing tasks in order of priority.

Accomplishment of the phased actions will be reported to the operations sergeant.

EXECUTION OF POMR

Phase 1. To be accomplished on Day 1.

_____ Offload and collect all ammunition, demolitions, and pyrotechnics for turn-in.

_____ Offload and account for all weapons and sensitive/serial numbered equipment.

_____ Top off vehicles from remaining fuel stores. If air movement is imminent, fill to 1/2 full.

_____ Inventory all nonexpendable BII.

_____ Thoroughly clean vehicles, both inside and outside.

_____ Conduct after operations PMCS on vehicles. Turn in DA Form 5988-E and DA Form 2404 to motor sergeant and parts clerk, retaining copy for the detachment's maintenance files.

_____ Initiate operator's corrective action.

_____ Clean all BII and OVM.

_____ Pay particular attention to cleaning the fuel cans.

_____ Clean all individual weapons and mounted weapons systems.

_____ Report completion of Phase 1 to operations sergeant. Phase I actions will be completed before the detachment is released.

Phase 2. To be accomplished on Day 2.

_____ Lubricate vehicles IAW lubrication order.

_____ Check on status of needed replacement parts through motor pool. Ensure any parts not on hand are on order.

_____ Initiate corrective actions, replacing and repairing parts. Upon completion, notify motor sergeant for entry on maintenance records.

_____ Order and replace detachment BII that were used, broken, or lost.

_____ Clean and rinse water cans.

_____ Give all weapons a second cleaning.

_____ Conduct after operations PMCS on all detachment equipment (other than vehicles) and take action as required.

_____ Repack all vehicles and vehicle equipment boxes.

_____ Report completion of Phase 2 to operations sergeant.

Phase 3. Complete all actions that could not be done during Phase 2 due to parts and/or equipment shortages. Give weapons a third cleaning. Submit any work orders on detachment during this phase. This phase may last until required parts and equipment become available.

OTHER REQUIREMENTS

The detachment—

- Keeps a copy of all maintenance records.
- Records and keeps copies of all job order requests.
- Maintains copies of requests for BII to know what items are on hand and on order.
- Cleans and inspects all weapons for three consecutive days.
- Dispatches vehicles, as required.

GARRISON MAINTENANCE REQUIREMENTS

The detachment conducts weekly PMCS on all vehicles (to include trailers) weekly unless precluded by events outside of its control.

- PMCS includes exercising the vehicles by at least driving the equipment around the motor pool and checking for proper operation.

- If the detachment cannot perform proper weekly maintenance, it ensures someone else (the motor pool or the company B-team) checks the equipment. This action is critical if the vehicles will be left sitting for longer than 30 days.

PMCS will be conducted IAW appropriate technical manuals and Group/Battalion maintenance SOPs.

Major maintenance events, such as semiannual services, lubrication orders, and major repairs will be planned IAW operational and training requirements. The detachment members conduct all services and repairs themselves, with minimal assistance from the battalion maintenance section. Such actions provide them experience that may be necessary when operating deep behind the FLOT without mechanic support.

Glossary

A/C	aircraft
AO	area of operations
AOB	advanced operational base
AP	armor piercing
AT	antitank
ATV	all-terrain vehicle
BII	basic issue items
CAS	close air support
CB	chemical and biological
CGU	cargo general utility
DA	Department of the Army; direct action
DKIE	Decon Kit Individual Equipment
DMA	Defense Mapping Agency
DMV	desert mobility vehicle
DMVS	Desert Mobility Vehicle System
DOM	desert operations motorcycle
DOT	desert operations trailer
DR	dead reckoning
DS2	Decontamination Solution 2
DZ	drop zone
FEBA	forward edge of the battle area
FID	foreign internal defense
FLOT	forward line of own troops
FM	field manual; frequency modulation
FOB	forward operational base
GAA	grease, automotive and artillery
gal	gallon
GL	grenade launcher
GMV	ground mobility vehicle
GPS	Global Positioning System
GW	guerrilla warfare
HALO	high altitude low opening
HF	high frequency
HB	heavy barrel
HMMWV	high mobility multipurpose wheeled vehicle
HTH	high test hypochlorite
IAW	in accordance with
IEDK	Individual Equipment Decon Kit
IR	infrared
km	kilometer
lb	pound
LBE	load-bearing equipment
LDS	lightweight decon system
LP	listening post
LRDG	Long Range Desert Group
LWCCS	Light Weight Camouflage Screening System
LZ	landing zone
MAC	Military Airlift Command

METT-TC	mission, enemy, terrain, troops, time available, and civilians
MG	machine gun
mi	mile
min	minute; minimum
MOPP	mission-oriented protective posture
MOS	military occupational specialty
MOOTW	military operations other than war
mpg	miles per gallon
mph	miles per hour
MRE	meal, read-to-eat
MSS	mission support site
MTP	mission training plan
NBC	nuclear, biological, and chemical
NBCC	nuclear, biological, and chemical center
NTC	National Training Center
NVG	night vision goggles
OEG	operational exposure guidance
OP	observation post
OPORD	operation order
OVM	operator vehicle maintenance
PDM	pursuit deterrent mine
PE	practical exercise
PLGR	precise lightweight Global Positioning System receiver
PLL	prescribed load list
PMCS	preventive maintenance checks and services
POI	program of instruction
POL	petroleum, oils, and lubricants
POMR	Post-Operations Maintenance and Recovery
psi	pounds per square inch
PSP	perforated steel planking
PVC	polyvinyl chloride pipe
PZ	pickup zone
RAD	remain all day
RT	receiver transmitter
SAS	Special Air Service
SDK	Skin Decon Kit
SF	Special Forces
SFG(A)	Special Forces Group (Airborne)
SFLE	Special Forces Liaison Element
SFODA	Special Forces operational detachment Alpha
SFOB	Special Forces operational base
SGT	sergeant
SO	special operations
SOF	special operations forces
SOP	standing operating procedure
SOT	special operations techniques
SP	starting point
SR	special reconnaissance
STB	supertropical bleach
T&EO	training and evaluation outline
TRP	target reference point

TTP	tactics, techniques, and procedures
UAR	unconventional assisted recovery
U.S.	United States
USAF	United States Air Force
USAJFKSWCS	United States Army John F. Kennedy Special Warfare Center and School
USGS	United States Geological Service
UW	unconventional warfare
UWOA	unconventional warfare operational area

REFERENCES

SOURCES USED
These are the sources quoted or paraphrased in this publication.

ARMY PUBLICATIONS
FM 3-3. *Chemical and Biological Contamination Avoidance.* (FMFM 11-17) 16 November 1992.
FM 3-4. *NBC Protection.* (FMFM 11-9) 29 May 1992.
FM 3-5. *NBC Decontamination.* (FMFM 11-10) 17 November 1993.
FM 3-6. *Field Behavior of NBC Agents (Including Smoke and Incendiaries).* (AFM 105-7; FMFM 7-11-H) 3 November 1986.
FM 3-9. *Potential Military Chemical/Biological Agents and Compounds.* (NAVFAC P-467; AFR 355-7) 12 December 1990.
FM 21-26. *Map Reading and Land Navigation.* 7 May 1993.
FM 55-450-4. *Multiservice Helicopter External Air Transport: Single-Point Rigging Procedures.* (FMFRP 5-31, Vol II; NWP 42-1, Vol II; AFR 50-16, Vol II; COMDTINST M13482.3) 11 February 1991.
FM 55-450-5. *Multiservice Helicopter External Air Transport: Dual Point Rigging Procedures.* (FMFRP 5-31, Vol III; NWP 42-1, Vol III; AFR 50-16, Vol III; COMDTINST M13482.4) 11 February 1991.

READINGS RECOMMENDED

These readings contain material used to get a historic overview of mounted desert special operations.

General

- *Bright Star in the Desert.* Unnamed author. Army Magazine, Vol 32, No. 2, February 1982.

- *A Desert War Might Leave US Forces High and Dry.* Frank Grove. Philadelphia Inquirer, November 15, 1981.

- *Special Forces for Desert Warfare.* John William Gordon, Jr. Unpublished dissertation, Duke University, North Carolina.

- *Desert Warfare.* Bryan Perrett. Thorson Publishing Group, Northamptonshire, England.

World War I

- *Light Car Patrols in the Libyan Desert.* Patrol after-action report from the archives of the Imperial War Museum, London, England. On file with ODA-512, 1/5th SFG(A).

- *Seven Pillars of Wisdom.* T.E. Lawrence. Doubleday and Co., New York.

- *Revolt In the Desert.* T.E. Lawrence. Doubleday and Co., New York.

- *Colonel Lawrence.* Liddell Hart. Dodd, Mead and Co., New York.

- *Lawrence.* Douglas Orgill. Ballantine Books, New York.

World War II

- *The Rommel Pages.* Liddell Hart. Harcourt, Brace and Co., New York.
- *The Phantom Major.* Virginia Cowles. Harper Brothers Publishing, New York.
- *War In the Desert.* James Lucas. Beaufort Books Inc., New York.
- *Handbook for the LRDG Mounted Officer in the Western Desert.* Ralph Bagnold. From the archives of the Imperial War Museum, London, England.
- *Providence their Guide.* MG D.L. Lloyd-Owen. The Nashville Battery Press, Nashville, Tennessee.
- *PoPski's Private War.* Col. V. Penoakof. Bantam Books, New York.

Oman Insurgency 1965- 1975

- *Operation Oman.* Tony Jeapes, SAS. The Battery Press, Nashville, Tennessee.
- *Where Soldiers Fear to Tread.* Randolph Fiennes. Hodder and Staughton, London, England.
- *We Won A War.* John Akehurst. Michael Russell Publishing, The Chantry, Wilton, Salisbury, England.

Yemen

- *Shifting Sands.* David Ledger. Peninsular Publishing.

Arab Israeli War

- *Sinai Victory.* S.L.A. Marshall. The Battery Press, Nashville, Tennessee.
- *The Yom Kippur War.* London Sunday Times, Doubleday and Co., Garden City, New York.
- *The War of Atonement.* Chaim Herzag. Little, Brown and Co., Boston, Massachusetts.

Algeria

- *The War Without a Name.* John Talbott. Faber and Faber, Boston, Massachusetts.
- *Wolves in the City.* Paul Henissart. Simon and Schuster. New York.

Gulf War

- *BRAVO Two Zero.* Andy McNab. Island Books, New York, New York 1993.
- *Military Lessons of the Gulf War.* Watson, Bruce W.; George, Bruce; Tsouras Peter. Greenhill Books, London 1993.

- *Lightning in the Storm, The 101st Air Assault Division in the Gulf War*. Taylor, Tomas. Hippocrene Books, New York 1994.

- *The Whirlwind War*. Kraus, Shubert and Theesa.. Center of Military History, U.S. Army, Washington D.C. 1995.

www.ingramcontent.com/pod-product-compliance
Lightning Source LLC
Chambersburg PA
CBHW050103230526
45470CB00004B/1657